AF342313

Philosophy of Engineering and Technology

Volume 22

The Philosophy of Engineering and Technology book series provides the multifaceted and rapidly growing discipline of philosophy of technology with a central overarching and integrative platform. Specifically it publishes edited volumes and monographs in: the phenomenology, anthropology and socio-politics of technology and engineering the emergent fields of the ontology and epistemology of artifacts, design, knowledge bases, and instrumentation engineering ethics and the ethics of specific technologies ranging from nuclear technologies to the converging nano-, bio-, information and cognitive technologies written from philosophical and practitioners perspectives and authored by philosophers and practitioners. The series also welcomes proposals that bring these fields together or advance philosophy of engineering and technology in other integrative ways. Proposals should include: A short synopsis of the work or the introduction chapter. The proposed Table of Contents The CV of the lead author(s). If available: one sample chapter. We aim to make a first decision within 1 month of submission. In case of a positive first decision the work will be provisionally contracted: the final decision about publication will depend upon the result of the anonymous peer review of the complete manuscript. We aim to have the complete work peer-reviewed within 3 months of submission. The series discourages the submission of manuscripts that contain reprints of previous published material and/or manuscripts that are below 150 pages / 75,000 words. For inquiries and submission of proposals authors can contact the editor-in-chief Pieter Vermaas via: p.e.vermaas@tudelft.nl, or contact one of the associate editors.

More information about this series at http://www.springer.com/series/8657

Colleen Murphy • Paolo Gardoni • Hassan Bashir
Charles E. Harris, Jr. • Eyad Masad
Editors

Engineering Ethics for a Globalized World

 Springer

Editors
Colleen Murphy
University of Illinois at Urbana-Champaign
Urbana, IL, USA

Paolo Gardoni
University of Illinois at Urbana-Champaign
Urbana, IL, USA

Hassan Bashir
Texas A&M University at Qatar
Doha, Qatar

Charles E. Harris, Jr.
Texas A&M University
College Station, TX, USA

Eyad Masad
Texas A&M University at Qatar
Doha, Qatar

ISSN 1879-7202　　　　　　　　ISSN 1879-7210　(electronic)
Philosophy of Engineering and Technology
ISBN 978-3-319-18259-9　　　　ISBN 978-3-319-18260-5　(eBook)
DOI 10.1007/978-3-319-18260-5

Library of Congress Control Number: 2015941149

Springer Cham Heidelberg New York Dordrecht London

Printed on acid-free paper

Springer International Publishing AG Switzerland is part of Springer Science+Business Media (www.springer.com)

Contents

About the Editors

Hassan Bashir is Associate Professor of Political Science at Texas A&M University at Qatar. His areas of research interest include political theory, non-Western political thought, and moral dimensions of globalization. His recent work focuses on inter-cultural political theory, history of political thought, Asian Politics, and non-state actors in international relations. Dr. Bashir is the author of *Europe and the Eastern Other: Comparative Perspectives on Politics, Religion and Culture before the Enlightenment* (Lexington Books, 2013). He is also the editor of *Co-existing in a Globalized World: Key Themes in Inter-professional Ethics* (Lexington Books, 2013) and *Global Citizenship and Multiculturalism: Political, Cultural and Ethical Perspectives* (Forthcoming, 2015). Dr. Bashir is the Founding Director of The Initiative in Professional Ethics at Texas A&M University, Qatar.

Paolo Gardoni is a Professor in the Department of Civil and Environmental Engineering at the University of Illinois at Urbana-Champaign. His research focuses on the development and use of probabilistic methods to model the impact of natural and human-made hazards on interdependent structural and infrastructural systems. He has been working on several multidisciplinary projects encompassing different areas including engineering, public policy, and ethics. Professor Gardoni is the Director of the Mid-America Earthquake (MAE) Center, established in 1997 by the National Science Foundation as one of three national earthquake engineering research centers. He is also the Associate Director of the Community Resilience Center of Excellence funded by the National Institute of Standards and Technology (NIST).

Charles E. (Ed) Harris, Jr. is a former Professor of Philosophy and former holder of the Sue and Harry Bovay Professorship in the History and Ethics of Professional Engineering at Texas A&M University. His specialty is applied and professional ethics. He and his co-authors have published a textbook in engineering ethics. He has also published a textbook in applied ethics and numerous articles in professional and applied ethics.

Eyad Masad is a Professor in the Zachry Department of Civil Engineering at Texas A&M University and the Mechanical Engineering Program at Texas A&M University at Qatar. He is a licensed professional engineer in the State of Texas. He teaches courses in engineering and ethics and has organized several workshops on professional ethics. He is also the Co-Founding Director of The Initiative in Professional Ethics at Texas A&M University at Qatar.

Colleen Murphy is an Associate Professor of Law and Philosophy and Director of the Women and Gender in Global Perspectives Program at the University of Illinois at Urbana-Champaign. Her research focuses on engineering ethics, the ethics of risk, and transitional justice in post-conflict societies. She is the author of *A Moral Theory of Political Reconciliation* (Cambridge University Press). Professor Murphy was a Laurance S. Rockefeller Visiting Faculty Fellow at the Princeton University Center for Human Values. Her research has received financial support from the National Science Foundation, the National Endowment for the Humanities, and the Qatar National Research Fund.

Chapter 1
Introduction

Colleen Murphy, Paolo Gardoni, Hassan Bashir, Charles E. Harris, Jr., and Eyad Masad

Abstract This volume considers the way(s) in which globalization complicates or alters discussions of engineering ethics and engineering ethics education. This introductory chapter lays out some of the key ethical questions that have emerged for engineers in their education, research, and practice due to the rise of a global economy and briefly summarizes each of the chapters in Parts I and II of this volume. The chapters in Part I provide an overview of particular dimensions of globalization and the criteria that an adequate engineering ethics framework must satisfy in a globalized world. The chapters in Part II consider pedagogical challenges and aims that arise in a globalized engineering ethics education curriculum.

Keywords Engineering ethics • Globalization • Engineering ethics education

1.1 Introduction

Engineering touches every facet of human life, from health and entertainment to food and the environment. Engineered products include the wheel, the pulley, and computers, as well as refrigerators, roads, and bridges. Engineers transform individual lives and the contours of communities, altering the form of all dimensions of our lives, including communication, transportation, and education through the artifacts and technology they produce. Frequently, lives change in ways unanticipated by engineers (Smith et al. 2013).

That engineering transforms lives through the artifacts and technology created is not surprising, since the aim of engineering is to respond to societal needs.

C. Murphy (✉) • P. Gardoni
University of Illinois at Urbana-Champaign, Urbana, IL, USA
e-mail: colleenm@illinois.edu; gardoni@illinois.edu

H. Bashir • E. Masad
Texas A&M University at Qatar, Doha, Qatar
e-mail: hassan.bashir@qatar.tamu.edu; eyad.masad@qatar.tamu.edu

C.E. Harris, Jr.
Texas A&M University, College Station, TX, USA
e-mail: e-harris@tamu.edu

© Springer International Publishing Switzerland 2015
C. Murphy et al. (eds.), *Engineering Ethics for a Globalized World*, Philosophy of Engineering and Technology 22, DOI 10.1007/978-3-319-18260-5_1

Engineering practice is designed to satisfy a particular goal or achieve a particular purpose: construct a bridge, design a new airbag, or retrofit a structure. An important standard of success for engineering products is functionality; that is, engineering products are successful insofar as they are useful for a given purpose. What is needed is a function of background circumstances, including economic conditions and the broader infrastructure. Consequently, the most technologically sophisticated artifact may not always be best suited to fulfill a particular need. A technologically less sophisticated artifact may be cheaper to produce and repair, and easier to use (Murphy et al. 2011).

Engineers contribute to the development of new technologies that influence and shape the way we live, in anticipated and unanticipated ways. For example, in 1935 when Wallace Carothers developed long, strong, and very elastic fibers, later called "nylons," he did not anticipate the widespread application of his work to consumer goods such as toothbrushes, fishing lines, and lingerie, or in special uses such as surgical thread, parachutes, or pipes. Nor did he anticipate the powerful effect nylons would have in launching a whole era of synthetics. More recently, in 1980 Tim Berners-Lee, a consultant at CERN (the European Laboratory for Particle Physics), wrote a program to link computers; he did not foresee that his program was the beginning of the World Wide Web, which in 2000 had more than sixteen billion Web user sessions (Nissenbaum 2001, 2004). The societal implications and impact of a given technology may be much greater than originally imagined and may vary across societies.

As this discussion highlights, a successful engineering solution cannot be simply technically sound, but also must account for cultural, social and religious constraints. Often the solution space is not well defined, making it harder to identify all possible solutions. Moreover, there is no single criterion that determines the best solution, and consequently it is frequently difficult to rank solutions. In practice multiple criteria, or external constraints, are used to evaluate possible engineering solutions. Cost effectiveness, ease of manufacture, safety, sustainability, and aesthetic appeal are just some of the more common criteria. Such constraints provide resources for delimiting possible solutions and guidelines for selecting among possible options, but because criteria are multiple, engineers must determine the relative weight that will be given to diverse kinds of considerations. Value judgments underpin both the criteria themselves and the relative weight assigned; using a given criterion, such as safety, to evaluate possible solutions reflects the judgment that safety is important. Placing greater priority on cost over aesthetic appeal reflects a judgment of the relative value of each kind of consideration.

Engineers are not only designers; academic engineers serve as advisors to graduate students. Academic and nonacademic engineers serve on professional engineering society committees and local agencies involved in developing and writing codes, designing guidelines, city standards, and specifications. Such codes, guidelines, standards, and specifications affect the entire society. Engineers also work for companies, and increasingly serve in managerial roles in this context.

Ethics is concerned with understanding what is good and of value in human life, and what constitutes the standards for right conduct for individuals and for justice in institutions. Engineering ethics considers these issues in the specific context of engineering. Questions considered include the standards for laboratory safety, fabrication, plagiarism, and authorship. Engineering ethics explores many issues in design and research. What standards determine when a system is safe enough to be used? When is the likelihood of a catastrophic event small enough to be ignored? At what point should an engineer be confident that all unforeseeable events and their implications are accounted for in the design? Engineering ethics also addresses issues of professional responsibility, professional advice, ethics support, and corporate ethics. Engineering ethics education is designed to help educate engineers to be cognizant of and take into account the social and ethical dimensions and implications of their work, instilling in students the importance and non-negotiability of recognizing certain ethical considerations inherent in engineering and of maintaining ethical standards. Successfully achieving these educational objectives is especially challenging in the context of the ongoing wave of globalization.

This volume considers the way(s) in which globalization complicates or alters discussions of engineering ethics and engineering ethics education. Globalization, which is essentially increased global interdependence, has fundamentally transformed the environment in which engineers learn and practice. There is an unprecedented demand today for engineers and other science and technology professionals with advanced degrees, due both to the offshoring of Western jobs and the rapid development of non-Western countries. In this new environment professionals in the high-tech fields are often required to work as part of international teams and devise solutions that are then implemented across national and cultural boundaries (Stukalina 2007). Motivated by the promise of mutual economic benefits, Western and non-Western countries are increasingly adopting strategies that encourage Western companies to establish foreign operations. In many cases, the costs of doing business in non-Western countries are low because of the compromises made regarding potentially harmful human and environmental effects. Not just the practice but also the teaching of engineering has gone international, and several Western universities and colleges have opened campuses in non-Western countries. As discussed later in this volume, the biggest challenge faced by these Western institutions is the diverse nature of the student body in terms of national origin, cultural orientation, and preparedness for entering higher education.

These changes have potentially far-reaching implications for engineering ethics. In the United States, professional engineering licensure is established as a standard measure of competency. However, most developing countries do not have a system for professional engineering licensing. This poses a challenge for international companies and local organizations in establishing a standard measure of competency for their engineers. Based on the experience in the State of Qatar and the surrounding region, some countries are requiring engineers to have professional licensing from the United States and/or European countries. Other countries are looking to establish their own professional exams and licensing. The interest in professional registration in engineering is becoming increasingly evident in various parts of the world, including Australia, Brazil, China, and Japan. In the United States, all Professional Engineer (PE) licensing is based on three requirements: education, experience, and examination. Education and experience standards are somewhat similar around the world, but examination requirements vary more widely. Western Europe also has a well-developed system for professional registration, usually based on a 5-year course of engineering education and an examination. Often delegations are sent to the United States, especially to the National Council of Examiners for Engineering and Surveying (NCEES), to which all state registration boards in the United States belong and which serves as a clearinghouse for information about registration in the United States.

With the increasing interest in mutual recognition of engineering licensure, an increasing interest in uniform standards can be expected to follow (Weil 1998; Love and Russon 2004). However, local traditions in custom and morality vary considerably. In some cultures, giving gifts is a way of cementing friendships. What looks like a bribe in a Western country might not be considered a bribe in other countries. Similarly, while nepotism is viewed negatively in most Western countries, in some parts of the world getting jobs for extended family members may be a very strong moral obligation. Thus in any attempt to formulate a model international code, universal requirements must be balanced against local traditions and cultural identities (Harris et al. 2009).

The current flow of technology and professionals is from the West to the rest of the world. Professional practices followed by Western (or Western-trained) engineers are, however, often based on presuppositions that can be in fundamental disagreement with the viewpoints of non-Westerners. Values and design constraints are varied, not only within individual communities but also across communities. If one does not take these educational and cultural differences into account, students may misunderstand the whole purpose of introducing students to ethical issues and moral decision-making. The lag between rapid technological advancement and the development of appropriate professional and ethical standards can lead to serious problems that may eventually nullify any gains from increased participation in the world economy for previously disadvantaged countries.

The chapters in this volume discuss some of the key ethical issues that have emerged for engineers in their education, research, and practice due to the rise of a global economy. Many of the chapters were originally presented at two conferences focusing on "Engineering Ethics for a Globalized World." The first conference was hosted at Texas A&M University in Doha, Qatar, in October 2011. The second was held at the University of Illinois at Urbana-Champaign in Champaign, Illinois in October 2012. The National Science Foundation in the United States and the Qatar National Research Fund both provided support for these conferences.

The authors, scholars and practitioners from diverse national, disciplinary and professional backgrounds discuss the ethical issues emerging from the inherent symbiotic relationship between the engineering profession and globalization. Their discussions facilitate a deeper and more complete understanding of the precise ways in which globalization impacts the formulation and justification of ethical standards in engineering, as well as the curriculum and pedagogy of engineering ethics education. Part I provides an overview of particular dimensions of globalization and the criteria that an adequate engineering ethics framework must satisfy in a globalized world. Part II considers pedagogical challenges and aims that arise in a globalized engineering ethics education curriculum.

Understanding the implications of globalization for engineering ethics requires first understanding what is changing as a result of globalization. In the first chapter, "Firms, Nations and Engineers: Considering Ethics in the New Global Environment" Leonard Lynn and Hal Salzman concentrate on the globalization of technology and scientific knowledge and provide an overview of important changes in how multinational enterprises (MNEs) operate. Commercial engineering takes place primarily through MNEs. Lynn and Salzman highlight structural changes in technology development, which is no longer the sole domain of "advanced" economies but increasingly occurs in emerging economies. Such changes challenge the assumption that advanced economies possess an innovation advantage. The increasing percentage of international students in science and technology programs has altered U.S. MNEs. In particular, management positions are held by an increasing number of these graduates. Occupants of decision-making positions thus have "the experience, familiarity, and linkages to facilitate the location of science and technology work globally, particularly to their countries of origin." Finally, outsourcing has expanded to include "high value-added functions." One important implication of these changes is that it is no longer obvious, for a number of reasons that Lynn and Salzman discuss, that MNEs can or should strive to give precedence to the home country's national interests. Lynn and Salzman consider some of the public policy implications of removing the assumption that a home country's interests should dominate. In particular, they argue that it behooves advanced countries to adopt a "collaborative advantage" model that seeks to identify mutual-gain strategies. "The goal for U.S. policy makers would be to benefit U.S. citizens through giving them a smaller piece of a much bigger pie, much as free trade policies historically have been far more beneficial to most people than mercantilist policies."

Norb Delatte's chapter "International Ethics and Failures: Case Studies" provides a comparative survey of how five different countries in both advanced and emerging economies have dealt with engineering failures. These case studies are the Hyatt Regency Walkway Collapse (1981) in the United States, the Malpasset Dam (1959) in France, the Vaiont Dam Landslide (1963) in Italy, the Sampoong Superstore (1995) in the Republic of Korea, and the Rana Plaza Building (2013) in Bangladesh. The specific focus of Delatte's analysis is on how engineers were held responsible by the legal system or professional societies. Different forms of legal sanctions include imprisonment or penalties. These cases highlight the very different approaches taken toward public safety in different countries, and the views of the role of engineers in achieving this aim. Delatte's discussion also draws attention to the importance of considering the source of a risk, or how a risk is created, in the process of risk management (Murphy and Gardoni 2011; Gardoni and Murphy 2013).

The final piece of important background context is provided by Rachelle Hollander, who in "US Engineering Ethics and its Connections to International Activity" provides a historical overview of engineering ethics, with a particular emphasis on the questions motivating work in the field. Hollander highlights how increasing complexity in the structure of organizations within which engineers work, often as a result of globalization, complicates questions of individual and organizational responsibility. Hollander concludes with a helpful summary of ongoing international efforts to set standards that affect engineering practice by the U.S. government, U.S. professional engineering societies, and the U.S. National Academies (NA), including NA efforts that involve similar bodies in other countries.

Hollander highlights a divide among scholars, who disagree about whether additional complexity requires new ethical concepts and/or new policies. Michael Davis directly addresses this issue, arguing in "'Global Engineering Ethics': Re-inventing the Wheel?" that it is a mistake to think that there are not already sufficient resources for dealing with the ethical dimensions of engineering in a global context. Davis argues that engineering is already global in the sense that engineers tend to share a common "culture," or distinctive terminology and technology. Specifically, Davis maintains that there already exist satisfactory global standards for engineering, including a global code of engineering ethics. Moreover, he claims, there is already an adequate global curriculum for engineering ethics. Finally, Davis cautions against the necessity and value of attempting to establish global registration or licensing requirements for engineering. Such efforts, he argues, would not contribute to the professionalism of engineers.

By contrast, subsequent chapters take as their point of departure the assumption that there are not sufficient ethical standards and education appropriate for engineers in the global context. In "Engineering Decisions in a Global Context and Social Choice," Noreen Sugrue and Tim McCarthy address the issue of how professionals should adjudicate among competing ethical standards in a global setting. Sugrue and McCarthy discuss three kinds of normative constraints that engineers must consider in their work—the values of the host country in which an engineer works, the values of the society from which engineers come, and the professional norms of engineering. They argue that adjudication among these constraints is best done by utilizing the Rawlsian original position construction, according to which an individual determines how the normative constraints would be balanced in a particular case by considering what an ideal rational agent would choose if he or she did not know his or her society or role. Sugrue and McCarthy illustrate the proposed methodology with three case studies.

A number of authors examine the basic concepts at the core of any ethical framework, considering what is the appropriate way to conceptualize these concepts in a global context. Charles E. Harris, Jr. in "Engineering Responsibility for Well-being" highlights the prominent role that a concern for well-being plays in engineering codes. He proposes an interpretation of the conception of well being that draws on Martha Nussbaum's capability approach. After discussing the implications of engineering for capabilities, Harris argues that promoting well being so conceived requires engineers to cultivate certain virtues. Harris ends by discussing the different virtues engineers need when they are designing in developing and developed nations. Developing societies should prioritize the virtues of empathy and compassion, while developed countries should emphasize a concern for the environment; sensitivity to the effects of technology, especially on society and human relationships; and creativity.

An adequate framework for engineering ethics must consider the relationship between engineering and development. In "Towards an Ethics of Technology and Human Development," Ilse Oosterlaken assumes that development ethics should be a part of any humanitarian engineering curriculum, and she proposes making a theoretical contribution to development ethics. Development is about making people's lives better and therefore about promoting well being. The ethics of technology and philosophy of technology rarely address technology in the context of poverty reduction or development in the global South, and development ethics that focus on global justice rarely mention technology. One area in which these scholarly endeavors could be joined is computer and information technology. Unfortunately there is relatively little work that combines information technology and development for underserved societies.

In order to explore and advance the ethics of technology and human development, Oosterlaken believes that it is necessary to investigate development ethics as it relates to engineers. This requires an exploration of the connections of engineering ethics to the ethics of technology and philosophy of technology. This can be done through a discussion of the application of the capability approach to technology, an approach that is already very influential within development ethics. Understanding the relation between technical artifacts and human capabilities requires us to move back and forth between a narrower ("zooming in") and a broader ("zooming out") approach. The narrower approach allows us to see the specific features or design details of technical artifacts, and the broader view allows us to see how technical artifacts are embedded in sociotechnical networks and practices. An example is the introduction of podcasting devices in rural Zimbabwe, which provide information about cattle management. From the narrower perspective attention was given to technology choice and the details of design; for example, the decision was made to use voice-based technology. Since many of the people were illiterate, a text-based technology would not lead to an expansion of human capabilities. A design change was later made to replace loudspeakers with headphones since it was determined that they would fit in with existing cultural practices. From the broader sociotechnical perspective, new cultural practices such as collective listening, discussion sessions, and demonstration meetings were encouraged. However, some of the broader sociotechnical aspects, such as the inability of the farmers to obtain some of the medicines recommended by the podcasts, could not be changed. While the project was not set up with the capability approach in mind, it did have an impact on the capabilities of the people. Furthermore, this impact depended both on the technical artifacts and their design, and on the broader sociotechnical networks in which the artifacts were embedded. Thus insights from STS and philosophy of technology were necessary in order to make the capability approach relevant to engineering.

Another central theme increasingly prominent in discussions of ethics, especially in a global context, is climate change. In addition, over the past 30 years there has been a significant attempt to think about the possible relation between the disciplines of ethics and economics. In "Ethics, Economics and the Environment" Khalid Mir focuses on how we should conceptualize the problem posed by climate change and the role of philosophy and economics in formulating this problem. Mir argues that engineers need to carefully consider the impact their decisions have on the environment and that the standard ethical approach used by economists in evaluating environmental damage—utilitarianism—is an insufficiently rich approach toward that end. This chapter begins by asking what motivates engineers and why, if at all, they should be concerned about ethics. Mir argues that engineers play a crucial role in shaping our world and the world in which future generations will live. Because of this, engineers must be attentive both to the ultimate aims of technology and to the various means devised for achieving them. Mir's key claim is that new forms of ethical thinking are required because of the circumstances created by our choices of technology and the lifestyles we have adopted. Mir also offers a critique of the utilitarian model of ethics in engineering, arguing that we should conceive of ourselves as part of a continuous moral community in which future generations are dependent on us, just as we were dependent on those before us. This approach creates a basis for emphasizing that each generation needs to be better stewards of the environment. In addition, if we conceive of future generations as dependent on us, we have a basis for our obligation to preserve, regenerate, and renew what we didn't create ourselves but simply inherited.

The final chapter of Part I concentrates on the construction industry as a case study. In "Ethics for Construction Engineers and Managers in a Globalized Market" George Wang and John Bruckeridge discuss the impact of globalization on the construction industry, illustrating some of the general dynamics noted by Lynn and Salzman. Wang and Bruckeridge focus specifically on the increasingly global scope of economic interaction. They note, for example, that managers of construction companies in the United States and Australia increasingly come from developing countries. More and more construction companies pursue projects in international markets. Foreign investors also buy or establish joint ventures with domestic companies. In many countries corruption is a problem for the construction industry, and Wang and Bruckeridge trace the differences in levels of corruption to differences in host country and sending country constraints, as well as to variations in professional norms. This is precisely the kind of variation Sugrue and McCarthy consider. Wang and Bruckeridge argue for the necessity of global professional registration that includes a professional ethics component.

The chapters in Part II concentrate on engineering ethics education. In "Overcoming the Challenges of Teaching Engineering Ethics in an International Context: A U.S. Perspective," Brock Barry and Joseph Herkert provide an overview of the ways globalization complicates engineering ethics teaching. The authors start from the assumption that teaching engineering ethics is challenging in countries such as the United States because many faculty members in engineering disciplines are uncomfortable teaching the material, due to their lack of credentials in the area. This discomfort is compounded in the international context. Against this background, the authors critically assess available instructional material, including videos, journals, textbooks, and online resources. One prominent limitation, they argue, is that current material is designed with a Western context in mind.

One chapter examines the study of engineering ethics in one specific non-Western context. Ruth I. Murrugarra and William A. Wallace consider Chile in

"A Cross Cultural Comparison of Engineering Ethics Education: Chile and United States." They discuss the adaptation of an ethics course, originally designed by and for a Western audience, to a Chilean classroom, with the purpose of assessing the impact of an educational experience emphasizing ethical behavior. To evaluate the efficacy of the course students were asked to fill out surveys during the first and final lectures of the class in order to gauge the students' perspectives on the relative importance of different values in life. For the Chilean course offering the course material was altered to provide either Spanish translations or alternate readings in Spanish. SIMULATE was used in English, but with clarifications provided for words/phrases/concepts that are specific to the U.S. setting. The responses from the surveys were analyzed for differences in nationality and ethnicity, and for the influence of team interaction on the surveys. American and Chilean students were found to hold the same values in similar importance, whereas immigrant groups differed vastly. The values of the students after taking the class were seen to change across all groups.

The final four chapters consider how engineering ethics education could be modified to better accommodate the globalization of the engineering profession. Research ethics is an area that is common to all engineering fields. In "Responsible Conduct of Research Training for Engineers: Adopting Research Ethics Training for Engineering Graduate Students," Sara Jordan and Phillip Gray argue that Responsible Conduct of Research (RCR) training provides a model for developing a global engineering ethics curriculum. In order to be relevant and truly global in character, engineering ethics should articulate universal standards, while taking into account differences in local contexts. They argue that RCR training integrates "near-universal" norms and provides a general framework for research practice that captures the commonalities among different fields of academic and professional study. Because it is research that unites all areas of inquiry, training in research ethics can serve as the foundational starting point for engineering ethics courses globally. Jordan and Gray include a case study of RCR training that occurred at a research university in Hong Kong, discussing the different perceptions of research integrity among postgraduate students in fields as varied as engineering and the social sciences. They conclude with recommendations for RCR courses that target engineers.

In "Training Engineers in Moral Imagination for Global Contexts" William Frey takes as his point of departure a different resource for global engineering ethics education: the moral imagination. Moral imagination is "the ability in particular circumstances to discover and evaluate possibilities not merely determined by that circumstance or limited by its operative mental models, or merely framed by a set of rules or rule-governed concerns." Referring to works by Charles Dickens, Frey warns against the pitfall of what he calls "telescopic philanthropy" when faced with ethical questions in the engineering profession. He then offers four guidelines for cultivating moral imagination in students: (i) recognizing the thinking of very different people; (ii) addressing foreign cultures from suitable points of view; (iii) using emotions like empathy, care, compassion, and hope; and (iv) realizing that good intentions are not enough. In teaching moral imagination, the author suggests three skill sets that engineering ethics programs should strive to cultivate:

an (i) ability to engage in role-playing, (ii) ability to recognize the different frames of a situation, and (iii) ability to engage in dramatic rehearsals. These skills help educators to prepare their students for the challenges posed by globalization. While moral imagination alone will not be sufficient to train engineers to work in global settings, engineers in multidisciplinary teams would be able to use the technique effectively. Moral imagination is especially useful in helping engineers become aware of the different responses people in different cultures have to situations and so to be more attuned to the increasingly "global" dimensions of their work. The essay focuses on engineering work in Puerto Rico, but has wide implications for teaching what some have called "techno-socio sensitivity" in the global context.

Sarah K.A. Pfatteicher's chapter "Sifting, Winnowing, and Scaffolding: Structured Exploration for Engineering in a Modern World" takes up themes considered by Jordan and Gray as well as by Frey. She notes that globalization tends to emphasize uniformity and commonality, but the global context is characterized by important cultural and religious differences. To equip engineering students to adapt to our continuously changing and diverse world, Pfatteicher argues, engineering ethics education should teach students the skills of "sifting and winnowing." Specifically, engineering education should not focus exclusively or even primarily on giving students information about the content of codes or extant standards; rather, it should cultivate in students the ability to distinguish the good and the bad, or the ethical and unethical, in the many different and often unforeseeable situations they will confront in their lives. To illustrate the sort of engineering ethics program she has in mind, Pfatteicher provides examples of what such an education might look like at the module, course, and program levels.

In "Toward a Global Engineering Curriculum," Eugene Moriarty maintains that "globalization" refers to increasing economic, political, and social integration, motivated by capitalism and transnational corporations. The globalization process is often relatively free of moral considerations. Although globalization can produce beneficial consequences, it can be responsible for perpetuating poverty and social inequality, and promoting ecological degradation, militarism, and nondemocratic forces. Globalism, Moriarty argues, can provide a perspective that corrects the ill effects of globalization. Globalism is a perspective that embodies the idea that we share a fragile planet and that we must embrace a concern for the environment and the values of sustainability, social justice, and community, as opposed to mere individualist self-seeking. Commitment to the values of globalism requires moving from what the author calls "standard engineering" to what philosopher Albert Borgman calls "focal engineering." The Accreditation Board for Engineering and Technology (ABET) has recently introduced requirements that point engineering curricula in a more global direction, mandating that engineering students demonstrate skill in communication, ability to work in multidisciplinary teams, and the broad education necessary to understand the impact of engineering on society and the environment. These requirements promote the values Borgmann associates with focal engineering with its concern for the particular rather than the general, the community rather than the individual, and the environment. While the ethics of Standard Engineering is based largely on the admonition to "do no harm," the ethics of Focal Engineering is based on the admonition to "do good."

Moriarty notes that some innovative engineering curricula point toward this new emphasis on global engineering. The University of Rhode Island offers an International Engineering Program, a 5-year course of study that consists of a BS in engineering and a BA in a language. Baylor University requires a course that combines technical writing and engineering economics, and another course on technical entrepreneurship in a global economy. Global engineering would focus on such problems as world hunger, low education levels, and environmental destruction.

Acknowledgements The material in this book is based upon work supported by the National Science Foundation under Grant No. 1066174/1240693. Any opinions, findings, and conclusions or recommendations expressed in this material are those of the author(s) and do not necessarily reflect the views of the National Science Foundation. This Book was also made possible by the NPRP Grant 4-1195-6-032 from the Qatar National Research Fund (a member of Qatar Foundation). The statements made herein are solely the responsibility of the author(s).

References

Gardoni, P., & Murphy, C. (2013). A scale of risk. *Risk Analysis*. doi:10.1111/risa.12150.

Harris, C. E., Pritchard, M., & Rabins, M. (2009). *Engineering ethics: Concepts and cases*. Belmont: Wadsworth Cengage Learning.

Love, A., & Russon, C. (2004). Evaluation standards in an international context. *New Directions for Evaluations, 104*, 5–14.

Murphy, C., & Gardoni, P. (2011). Evaluating the source of the risks associated with natural events. *Res Publica, 17*(2), 125–140.

Murphy, C., Gardoni, P., & Harris, C. E., Jr. (2011). Classification and moral evaluation of uncertainties in engineering modeling. *Science and Engineering Ethics, 17*(3), 553–570.

Nissenbaum, H. (2001). How computer systems embody values. *IEEE Computer, 120*, 118–119.

Nissenbaum, H. (2004). "Will security enhance trust online, or supplant it?". In P. Kramer & K. Cook (Eds.), *Trust and distrust within organizations: Emerging perspectives, enduring questions* (pp. 155–188). New York: Russell Sage Publications.

Smith, J., Gardoni, P., & Murphy, C. (2013). The responsibilities of engineers. *Science and Engineering Ethics*. doi:10.1007/s11948-013-9463-2. pp. 1–20.

Stukalina, Y. (2007). Globalisation and engineering education: Preparing students for professions of the 21st century in science and technology. *Transport and Telecommunication, 8*(1), 30–39.

Weil, V. (1998). Professional standards: Can they shape practice in an international context? *Science and Engineering Ethics, 4*, 303–314.

Part I
Ethical Issues in a Globalized World

Chapter 2
Engineers, Firms and Nations: Ethical Dilemmas in the New Global Environment

Leonard Lynn and Hal Salzman

Abstract Global firms—which are the primary organizational entities through which commercial engineering is conducted—have largely lost attachment to their country of origins. Although they may retain some cultural heritage that stems from their country of origin, they are increasingly global organizations that have multiple "host countries" in which they have located some portion of their operations. As a result, they have a decreasing degree of investment in, or loyalty to, any one nation. At the macro, policy level, this new globalization raises questions about the context in which engineering work (as well as other work) is conducted. This structural change in firms and the process of technology development raises questions about engineering values and ethics. This paper will provide an overview of the structural changes in technology development based on fieldwork conducted by the authors, and discuss the value and policy implications of these changes.

Keywords Nationalism • Technonationalism • Globalization • Collaborative advantage • Stakeholders

2.1 Introduction

"Stakeholder analysis" is a mainstay of engineering management education offered in business schools. According to this perspective, the firm has responsibilities that go beyond maximizing profits for its owners. These responsibilities extend to those who have interests in the firm's operations or its products. These other "stakeholders" include employees, customers, business partners, communities, future generations, the environment, the various levels of government, regulatory bodies, civic institutions, special interest groups, trade groups, and even competitors.

L. Lynn (⊠)
Case Western Reserve University, Cleveland, OH, USA
e-mail: leonard.lynn@case.edu

H. Salzman
Rutgers University, New Brunswick, NJ 08901, USA
e-mail: HSalzman@Rutgers.edu

© Springer International Publishing Switzerland 2015
C. Murphy et al. (eds.), *Engineering Ethics for a Globalized World*, Philosophy
of Engineering and Technology 22, DOI 10.1007/978-3-319-18260-5_2

Engineering ethics education often focuses on the multiple stakeholders and the tradeoffs practicing engineers need to consider in making decisions that might benefit their employer, but may also harm other stakeholders. A "stakeholder" that has received little explicit attention, but is often taken as a given, is the nation. While some, particularly in the policy-making community, assume that ethical behavior requires engineers and firms to be "patriotic," the relationship between patriotic behavior and ethical behavior in the context of technology development in today's environment has been given little thought.

This lacuna in stakeholder analysis is of increasing consequence as the practice of engineering becomes increasingly globalized. For example, the calls on engineers to support nationalistic interests in the United States have grown in the past decade. The influential report from the National Academies of Science, *Rising Above the Gathering Storm* (2007), to give one contemporary example, likens the technological and economic ascendance of emerging nations to Churchill's characterization of Germany's ascendance before World War II. By what measure is such a characterization invoked?

This report and many others like it have come to define national technology policy and an implied set of ethical and moral underpinnings akin to those at play in the nationalistic sentiments of the Second World War and Cold War periods. The U.S. President has called upon the nation to respond to the new science and technology "threats," saying we need "another Sputnik moment," as though the nation is in a technological race with countries that, like Nazi Germany or the Soviet Union (USSR), pose existential threats to the United States. Beyond the obvious moral and ethical differences overlooked in the rhetorical fever of these claims of equivalency between the current advances of technology in the emerging nations and technological advances by nations that posed direct military threats during the twentieth century, there are also questionable underlying assumptions about the national identity of today's multinational enterprises (MNEs). No longer can we assume that a nation's interests will be addressed through advancing the "competitiveness" of its companies.

Historically, the United States was assumed to be a key stakeholder in U.S.-headquartered MNEs. U.S. government social and economic policies and expenditures were directed to support these firms in everything from education and training funding to tax breaks, infrastructure, and even military defense of their assets around the globe. The return to the nation for these investments was, as these firms grew and prospered, their contribution to the technological and economic strength of the U.S. Following in that historical perspective, engineers are now called on in *Rising Above the Gathering Storm* (2007) and other prominent reports to help build the rampart to fortify the dominant global position of the U.S. But what happens when engineering workforces, firms, and educational institutions are globalized? Should engineering work and education be directed towards advancing a company's interest even when it may be at the cost of a perceived national interest? Or should engineers be expected to sacrifice corporate interests for perceived national interests? And, what if the nationality of the company and the engineer are not the same? What are the ethics and values of engineering in a new global environment?

In this chapter we challenge the assumption that "U.S." (or European, or Japanese, or Chinese, or Indian, or Brazilian) MNEs can be expected to accept as an ethical imperative the precedence of home country national interests over other interests. And, conversely, we raise the question of whether employees working for an MNE should be expected to give priority to the MNE's home country interests over those of the employee's home country or some other country of interest. In this regard, we argue that any normative claims on the firm based on nationality are largely imponderable in the current global environment, if not irreconcilable with competing normative claims (e.g., to provide benefits to emerging nations that have been providing talent through brain drains as well as natural resources inexpensively to advanced industrial countries).

As an approach to mitigate these quandaries, we suggest that more attention be given to the "collaborative advantages" that can come from a focus on positive- rather than zero-sum interactions as engineers work in cross-national environments. We suggest, as well, that those in government overseeing science and technology policy re-examine, in a global context, the ethics of some of their policies.[1]

2.2 Globalization of Engineering: A Convergence of Trends

Before examining some of the ethical implications of today's accelerating globalization of technology for engineers in multinational companies, it is useful to understand what lies behind this globalization, and what makes it different from past periods of globalization (Lynn and Salzman 2004). These new characteristics of globalization include important structural changes in the international distribution of innovation activities and human capital flows, as well as changes in corporate organizational forms, structures, and functioning. One consequence has been an "innovation shift" in which pioneering technology development is no longer the exclusive preserve of "advanced" economies, but is more frequently occurring in emerging economies. This raises questions regarding longstanding views about the persistence of inherent innovation advantages of advanced industrial nations and of particular regions, such as Silicon Valley. Theories of the "geographical stickiness" of technology (e.g., von Hippel 1994; Porter 1998; Saxenian 2000) propose that

[1]The research this paper is based on and was supported through grants from the National Science Foundation (Human and Social Dynamics Program, #SES-0527584; Social Dimensions of Engineering, Science and Technology #0431755) and the Alfred P. Sloan Foundation. Additional support was provided by the Ewing Marion Kauffman Foundation to study technology entrepreneurship and globalization. Earlier versions of this paper were presented at "Engineering Ethics for a Globalized World" (EGW11 and EGW12) Conferences, Doha, Qatar, October 2011; University of Illinois at Urbana-Champaign, October, 2012; Organizers: Colleen Murphy, Paolo Gardoni, Hassan Bashir, Ed Harris and Eyad Masad. We also thank David Hersh for superb research and analysis of educational outcomes and Christine Jenter for diligent editing and research assistance.

some regions have a unique mix of firms, capital, culture, and talent that makes them spawning grounds for innovation. While it might be thought that these regions can give an enduring advantage to their host nations, there are strong reasons for doubting this in the current environment of globalization. To be sure, many of these regions are likely to remain powerful centers of innovation, but the emerging economies are also developing regional innovation clusters and industries of their own that, in some areas, will be on a par with those in the advanced industrial nations. Over the longer term it seems unlikely that any nation can permanently remain dominant in the creation of technology across the spectrum.

A second major development is that U.S. graduate schools and the U.S. S&T (science and technology) workforce have become internationalized over the past few decades. Students on temporary visas have generally made up 20–50 % of the graduates of U.S. science and engineering graduate programs since the late 1980s. Some programs, such as IT-related programs, experienced sharp spikes in the number of foreign student graduates in the late 1990s and, for most programs, there has been a constant or slowly increasing rate of foreign student enrollments over the past 20 years. Over this period, many of these graduates have entered U.S.-based firms. They now make up a significant proportion of the U.S. science and technology workforce. A number of these scientists and technologists have now moved into senior technical and middle- and upper-level management positions in U.S.-based firms. Now in decision-making positions within U.S. MNEs, and as business strategy develops to globalize firms, these managers and S&T workers have the experience, familiarity, and linkages to facilitate the location of science and technology work globally, particularly to their countries of origin. This is a point we will return to later.

A third major development has been a change in the structures of major multinational firms. Historically, these firms tended toward ever-greater integration of all parts of their production and services systems, first through horizontal, then vertical integration, then to multidivisional expansion. This led to growth in organizational size and in the scope of activities and functions. Firms also were firmly rooted in their "home" geographies, which aligned a firm's economic performance with that of the nation in which it was based. R&D, as a strategic competency, was kept close to company headquarters. During the late 1980s this began to change. Outsourcing began as large manufacturing firms started buying rather than making commodity parts. Firms gradually expanded the scope of outsourcing to the external acquisition of innovation and high value-added functions. This change in innovation strategy occurred throughout many industries and, in a remarkable shift, Wall Street came to consider firms to be poorly managed if they relied on strong internal R&D rather than the external sourcing of technology (Lynn and Salzman 2007). This change in organizational form provided critical elements in the foundation for the globalization of science and technology work we are now witnessing. U.S. MNEs profited from this change which lowered their costs and increased the pool of human resources they can access. Further, the international workforce of U.S.-based MNEs facilitated this globalization by providing these firms with crucial cross-cultural experience and knowledge. Indeed, the more integrated organizational form and less international workforces of European and Japanese firms have caused them to be somewhat slower than their U.S. counterparts in their globalization, especially of high-level activities (e.g., see Lynn and Salzman 2007; Lynn et al. 2012).

All of these changes also have provided new advantages to emerging economies, giving some of them world-class capabilities when it comes to technological innovation. In IT products and services, the initial offshoring of low-level activity, such as IT remediation programs in the 1990s to address anticipated Y2K problems, led to offshore companies implementing highly structured and systematized methods of software development. As IT technologies mature, the innovation shifts from product to process development, which can lead to more reliable and secure software. Also changing are the types of innovation developed within the local context of the emerging economies. In previous stages of globalization, local innovation was confined to adapting existing products to local conditions. Now, MNEs are increasingly finding innovations conceived for local environments may have global applications. Additionally, innovation is increasingly occurring in both high-end and low-end technology.

In the past, it was typically only high-end innovation that pushed the technology frontier. Now, low-end innovation may provide opportunities for new technology development and high profit. For example, although premium smartphones are a high-growth area (7.3 billion units globally in 2014),[2] the middle and bottom of the global income pyramid are a far larger, and a largely untapped market for low-cost smartphones. Just one cell phone provider in China has over 800 million subscribers, growing at two to three million per month,[3] as compared to the largest U.S. carrier, Verizon, with 124 million subscribers and growing at approximately 585,000 subscribers a month.[4] At over 300 million cell phones in use in the U.S., or 103 phones per 100 people, the market would seem close to being saturated (and thus relying on product obsolescence for sales growth) as compared to China's 93 phones per 100 people, or India's 77 cell phones per 100 people, each with a population of nearly one billion and growing.[5] Although there is global growth for expensive smartphones, there is a far larger market for the company that innovates further down the price chain. The smartphone is just one example of the innovation technology demand from the emerging countries and mid-income market segments.

It is not just that the largest markets for technology and innovation are outside advanced industry economies and thus at sites that are increasingly likely to spur

---

[2] https://gsmaintelligence.com Retrieved February 18, 2015.

[3] http://www.chinamobileltd.com/en/ir/operation.php?section=number&year=2014 China Mobile, Investor Relations, 2014.

[4] http://www.fiercewireless.com/special-reports/grading-top-8-us-wireless-carriers-third-quarter-2014?confirmation=123 Retrieved February 18, 2015.

[5] http://en.wikipedia.org/wiki/List_of_countries_by_number_of_mobile_phones_in_us Retrieved February 18, 2015.

innovation different from that developed in high-income markets, but also that technology flows across boundaries. Innovations developed in the home country of an MNE may be quickly transferred to locations elsewhere in the world. In today's environment innovation leadership by an MNE does not necessarily confer greater advantage to the MNE's nominal home country than to the other countries in which it operates.

In a series of research projects we interviewed engineers and engineering managers at more than 75 sites of 46 multinational and entrepreneurial firms. These included U.S., European, and Japanese MNEs in software, auto parts, electronics and electrical equipment at sites in the U.S., China, India, Mexico, Japan, Germany, and the U.K. as well as entrepreneurial firms working on technology development with the MNEs (Lynn and Salzman 2007; Lynn et al. 2012). We found a common pattern in the evolution of globally distributed engineering projects. First, firms typically begin by locating lower-level work at their offshore site, but as these sites develop their own capacity—hiring and training more educated and skilled workers, attracting emigrants to return, and gaining experience and confidence— "engineering creep" occurs. By this we mean the offshore site expands the range of work it does to include higher and higher level engineering tasks. Sometimes this complements what is being done in the firm's home country sites; other times it replaces it. This progression up the "innovation value chain" is a newly developing phenomenon, and we do not see any indication that there are inherent limits to the level of activity that can occur in emerging countries. Human capital is becoming ever more available, and financial capital is available as well. The large potential markets in China, India, Brazil, and elsewhere lead firms to justify investments even for expensive labs and development facilities in these countries (for examples of this see Lynn et al. 2012).

2.3 Company and Country: A Dissolving Link?

The Chairman of the Board of General Motors in the early 1950s, Charles Erwin Wilson, has been famously misquoted as saying, "What's good for General Motors is good for the United States." What he actually said during confirmation hearings when he was nominated by President Eisenhower to serve as U.S. Secretary of Defense was that he could not conceive of a situation where a decision made by the Secretary of Defense would be adverse to the interests of GM, "because for years I thought what was good for our country was good for General Motors, and vice versa." Although, given subsequent discussions of how GM managers in Europe (along with managers from IBM and other major U.S. firms) cooperated with the Nazis in the years leading up to World War II this may have been misleading at best, in the 1950s there was, nevertheless, a clear sense that having strong "U.S. firms" benefitted the U.S., and that U.S. firms benefitted from being citizens in the country that dominated the world economically, technologically, and militarily. Based on this notion, over the years U.S. Federal and State governments adopted "buy American" policies and, for many years, "buying American" was viewed as patriotic.[6] Nor, of course, was this attitude unique to the U.S. European governments supported companies such as British Steel, Olivetti, CL, Bull, and others in the hopes that these firms would become "national champions" beneficial to the home country by reinforcing its technological strength in strategic industries. It is noteworthy that these ventures have not been particularly successful.

For that matter, the notion that firms even have a nationality seems increasingly problematic in today's global economy. In recent years firms like Chrysler, Nissan, Sony, and Pepsi have had foreign CEOs, and many others have had foreign members on their boards. As we noted, middle management at MNEs also tends to include many foreign or immigrant employees (perhaps a third of the engineering managers we interviewed at U.S. multinationals were not American by birth). Major firms from most countries have foreign bond and shareholders. Most have offshore operations and large numbers of foreign employees. And the days when an MNE's wellbeing was almost totally dependent on domestic sales are long gone.

At a Chinese facility of a U.S. MNE, the engineering management team we interviewed included a Chinese-American, two mainland Chinese, and a Chinese from Taiwan. In discussing their career histories and aspirations it became clear that these engineering managers identified much more with the MNE they worked for than with China, Taiwan, or the United States. A Japanese engineering manager we interviewed at a U.S. multinational had previously worked at Japanese and German multinationals. He saw little significance in the nationality of the firms. In some cases Chinese and Indian engineers who had risen high in U.S. firms strongly promoted the establishment of operations in their country of origin, with the hope of returning to their homeland. Some expressed hopes that this would benefit their native country. Understandably, these managers did not find "America first" or other U.S. nationalist objectives as integral to their personal or corporate mission. Certainly they did not share the nationalist sentiments such as those expressed in the *Rising Above the Gathering Storm* report. That report identified China and India as threats to the U.S. or enemies to be quashed. Framing policy and national objectives as zero-sum is unlikely to find sympathy in the growing, international strata of S&T workers, managers, and executives. However, win-win scenarios for nations may provide a common strategic path.

---

[6]In the late 1990s there was a shift in the location of IT work, moving the more routine and lower-skilled work offshore and using lower-cost offshore firms to do the service work onshore. There was widespread political reaction. In the course of a single year, 2004, the legislatures in 40 states introduced a total of more than 200 bills restricting offshoring (compared with legislation proposed in only four states the year before; see Salzman 2013). And the presidential candidate, John Kerry, in a speech to his supporters, denounced offshoring firms and promised to eliminate tax loopholes for any "Benedict Arnold company or CEO who take the jobs and money overseas and stick you with the bill." Recently there seems to be a resurgence in the equation of buying American with patriotism, at least in many advertising campaigns.

Finally, over the past decades corporate strategy and policy interests suggest other limits to the strength of U.S. MNE loyalty to, or investment in the economy of the United States. For example, while in a bid to increase visa caps, a number of high-tech CEOs discussed the demand their companies had for U.S.-based science and engineering workers to a *Wall Street Journal* reporter in June, 2006. One was quoted as follows:

> Craig Barrett, chairman of Intel Corporation, says his company employs most of its researchers in the U.S. and wants to keep it that way. The reasons? ... "If engineering is happening here in the U.S., I think my children will have a richer work environment" (*Wall Street Journal* 2006).

However, during the previous year, another high-tech company, Sun Microsystems, made the following announcement to Wall Street analysts:

> Sun Microsystems Inc. has chosen four of its facilities around the world *to take the place of its Silicon Valley office as the research and development hub* .... "We are over-invested in high-cost geographies like the U.S., and underinvested in low-cost geographies like India," ... the company's senior vice president of global engineering told reporters in Bangalore. (Associated Press 2005; emphasis added).

However, even the move to low-cost geographies didn't save the company and by 2010 it was acquired by Oracle and many of its employees laid off (Perry 2014). Over the past decade and a half, another technology giant, IBM, has reduced its U.S. IT workforce by 30 % and now has four times more offshore than U.S. employees. But that shift appears to have been insufficient and another reduction in the U.S. workforce of perhaps 26 % is planned (Perry 2015).

At a meeting in 2011 Steve Jobs told President Obama that Apple would have located 700,000 manufacturing jobs in the U.S. instead of China, but was unable to find enough U.S. engineers to support its operations. Perhaps, but it might be noted that the U.S. national average wage for electronics production workers was some $42,000 compared to the $4,800 a year paid in China by Apple's contract manufacturer. The difference between U.S. and Chinese wages multiplied by 700,000 comes to about $26 billion, slightly more than Apple's net profit for 2011. How strongly would Apple shareholders have supported Apple's attempt to be patriotic? Would they have considered the sacrifice in *their earnings* in the greater U.S. national interest? What about non-American shareholders?

Another striking example of ethical dilemmas of global business strategy and investing in the firm's country of origin is that of multinational Emerson Electric, a company celebrating the 125th year since its founding in St. Louis, MO in 1890. Now headquartered in Ferguson, MO, the CEO of this *Fortune 500* company, David Farr, stated shortly after taking over the helm of the company in 2000 that he was shifting jobs at all levels out of the U.S. As this strategy was reported in 2002:

> For more than a century, Emerson Electric Co. rode the business cycle, expanding when sales were high and cutting jobs when revenue went south.
>
> That is still the strategy as the U.S. economy is poised to turn the corner, but there is a twist: The St. Louis-based multinational won't be hiring at its rural Mississippi plant—that facility is shutting down. Instead, it expects to be hiring in China, the Philippines or Mexico.

And it won't just be assembly-line jobs going offshore. Emerson hopes to move at least half of its engineering work to China and India (Iritani 2002).

Farr's offshore strategy was reported again in 2004, quoting him as saying "If half of your sales go outside of the United States, you're going to have half of your engineering outside of the United States, too," (Wiggin 2010), and then again in 2009, saying "I'm not going to hire anybody in the United States. I'm moving" (Gallagher 2011).[7] Two years later the *St. Louis Post-Dispatch* named Farr "St. Louis Citizen of the Year" for his charitable work in the community even while noting his statements, record, and business strategy of moving jobs offshore and decreasing his U.S. hiring (Gallagher 2011). In the first part of 2015, half of the job openings at the company were for positions outside of the U.S.[8]

The U.S. CEO of a U.S. MNE may well have patriotic aspirations, but the pressures of the market can be even stronger. Intel, despite whatever intentions it may have expressed under a previous CEO (Barrett ended his term as Chairman of the Board at Intel in 2009), has now followed other firms in globalizing many of its operations. To be sure, Intel and a few other firms now seem to be returning some activities to the U.S., but this is because of changes in relative costs, not selfless patriotism. And, as the case of Emerson's CEO Farr illustrates, patriotism, or at least model citizenship, does not appear to pose ethical or value conflicts with also pursuing business strategies that lead to moving employment opportunities from the firm's community of a 125 years to communities elsewhere in the world where taxes and wages are lower, where government policies seem more favorable, or where there are higher growth markets. The potential ethical and value conflicts between, for example, honoring a firm's charity to address problems of unemployment and homelessness while decreasing the employment opportunities in its community were only briefly part of the public discussion when the offshoring movement was first growing a decade ago, as when presidential aspirant John Kerry denounced offshoring firms and promised to eliminate tax loopholes for any "Benedict Arnold

---

[7] In 2009, however, this statement was issued as a threat reflecting his displeasure with President's Obama's policies (though Farr also had described this as his business strategy years earlier during the Bush presidency). The statement was reported coming "… after complaining that the policies of President Barack Obama's administration threatened business, he said he was 'not going to hire anybody in the United States. I'm moving.' He now calls that an exaggeration, noting that he's hired Americans since. But he doesn't back off the message.

Investors who held Emerson stock when Farr took over in 2000 have seen a 112 % total return, versus 18 % for the S&P 500 Index of large-company stocks. That's what counts to Wall Street. 'He's very good, and well regarded in the investment community,' says Rich Kwas, analyst at Wells Fargo Securities in Baltimore. Kwas credits Farr for deftly managing the company through the Great Recession, in part by 'moving head count to low-cost countries,' he says." (Gallagher 2011).

Interestingly, in the reports of these comments in 2009, none mentioned that he had articulated the same strategy earlier, under President Bush and that it was fundamentally a business strategy seemingly independent of specific presidential policies in any particular moment.

[8] Of the 1,236 job openings listed on the Emerson careers website, 633 were listed as located in the U.S. http://emerson.jobs/ Retrieved 22 February 2015.

company or C.E.O. who take the jobs and money overseas and sticks you with the bill" (Salzman 2013; 60).

It should be further noted that it is not just the United States (or for that matter the U.S., Japan, EU, and other advanced economies) that have MNEs. Recent research by the McKinsey Global Institute shows that the share of *Fortune Global 500* companies from emerging economies will probably jump to more than 45 % by 2025, compared to only 5 % in 2000. McKinsey concludes that the new MNEs could disrupt entire industries "by designing superior products, by bringing them to market faster, and by streamlining business processes" (McKinsey Global Institute 2013). Promoting national champion strategies is likely to become less and less advantageous to the U.S. as non-U.S. MNEs become more and more prominent. Indeed, the "national champions" given special government support in Japan and Europe in the computer, software, and other industries did little more than waste public resources. It is thus a strategic as well as an ethical imperative to develop policies that return benefit to the nation, but are not specifically targeted at dominating industries at the expense of other nations.

2.4 Nations in Competition: "King of the Hill" Technology Strategies

In an increasingly globalized world a common, and we believe misguided, aspiration is for the policymaker's country to use technology to achieve and maintain a kind of technological hegemony at the "top of the value chain" (e.g., NCEE 2007). The ideal scenario according to this view, and implicitly a goal of many policy proposals in the U.S., would be one in which most of the U.S. workforce is employed in "creative work" with low-skilled jobs either located in emerging economies or performed by machines. Leaving aside for the moment the ethical stance suggested by such aspirations, this future world seems improbable.

First, as we have noted, appeals to national interests are less and less likely to appeal to MNEs that are fast losing their sense of national identity. Second, one wonders how the U.S., or any other country, would manage to muster the human resources needed to climb to the top of all of the most crucial technology value chains and stay there. Nor is it clear why this would even be desirable for most Americans. Setting these points aside for the moment, let us look at the specific policies advocates of the top-of-the-value-chain model are proposing. The thrust of their proposals is based on an assumption that the U.S. is losing its technological pre-eminence because it does not have a sufficient number of science, technology, and engineering workers, and that this shortfall is forcing firms to go offshore to meet their needs. Thus, they propose that the U.S. improve its K-12 math science education and that it take measures to entice more young Americans to enter engineering programs at universities. They also propose visa programs that encourage more foreigners to come to the U.S. to work.[9]

But what would happen if these policies actually worked, and more Americans entered science and engineering programs at universities? A major question is whether or not these new graduates would actually end up with Science, Technology, Engineering, and Math (STEM) jobs. Even now only about a third to half of new STEM graduates in the U.S. actually move into STEM jobs.[10] And many of those don't remain in STEM very long. For the 1993–2001 cohorts of STEM graduates who took STEM jobs nearly one half were no longer in STEM jobs 2 years after graduation. Another 20 % of the STEM graduates were still in school but not in STEM majors (NSF 2006, table 3).[11] Simply put, the market for STEM graduates is not very attractive. Other jobs often offer higher pay. Moreover, careful academic studies, notably by Richard Freeman and colleagues (2004, 2006), point to a comparative decline in STEM wages. Indeed, other research finds that not just comparative, but real wages in STEM occupations were declining in the 1980s and 1990s, well before the bust of the dot-com bubble and the 2008 global financial meltdown (Espenshade 1999; also see Stephan 2012). Interestingly, engineering managers we interviewed in previous research at engineering and technology firms did not claim a shortage of applicants, nor did they complain about the math and science skills of applicants. The skills deficits most often mentioned were the "soft skills" of communication and ease in working across organizational, cultural, and disciplinary boundaries (Lynn and Salzman 2002; Salzman 2000).

In brief, it seems that if the U.S. does not have enough STEM workers to pursue a broad-based, top-of-the-technology-value-chain strategy it is *not* because the U.S. doesn't educate enough engineers (or because of weaknesses of the U.S. K-12 education system), it is because there are not enough STEM jobs to attract and retain Americans into these jobs. The problem, if it is a problem, is one of demand not supply. Non-market efforts to entice people into science and technology jobs would put more Americans into jobs they would not otherwise want to take—itself an ethical issue in a country valuing freedom of choice.

---

[9]The following sections draw on, and excerpt in part from, an analysis by Lowell and Salzman (2007), Salzman (2013), and Salzman et al. (2013).

[10]From 1985 to 2000, around 450,000 STEM degrees were awarded to U.S. citizens and permanent residents. Over the same period, STEM occupational employment increased about 150,000 annually.

[11]The data also fail to support the claim that U.S. students show declining interest in science and engineering fields, either in college or in entering the workforce. The actual numbers of STEM college graduates has increased over the past three decades and held steady in recent years. The "continuation rate" of STEM bachelor's graduates going on to graduate school, has also remained at a steady rate for the past two decades. The major change since the 1960s, of course, has been the large increase in foreign-born students (on temporary visas) entering graduate school and the workforce.

2.5   A World of Collaborative Advantage

Top-of-the-value-chain strategies to guide national policies, then, pose two sets of problems. On the one hand, developing an adequate supply of properly educated S&T workers *who actually want to pursue such careers* is problematic. Nor does recent experience suggest that "U.S. MNEs" are willing or able to make the financial investments necessary to locate most of their high-end jobs in the United States, that is, by paying salaries competitive with those that S&T graduates can get in non-S&T fields. They not only would have to give up savings from the current gap between emerging economy and U.S. salaries, but would have to sacrifice more by widening the gap when they increase the financial incentives for U.S. S&T workers to remain in S&T. And, of course, they would also face the potential loss of markets in emerging economies angered by U.S.-first policies.

Changes in U.S. visa policies to facilitate the movement of foreign S&T workers into the United States may offer some advantages in increasing the technological strength of the U.S., as we shall discuss shortly, but caution is needed in determining the exact purpose of such policies and crafting the policies to achieve those purposes. Policies to bring in entry-level S&T workers on a temporary basis, for example, may reduce the costs of some activities in the U.S., but they will also discourage Americans from pursuing such careers by depressing entry-level wages. Moreover, if foreign nations see the U.S. as attempting to strip them of their "best and brightest," there will likely be backlashes. And what of the ethical sentiments of non-American S&T people who are pressured to favor nationalistic American interests that may conflict with those of their home country? Finally, if foreign S&T workers in the U.S. decide to return to their country of origin, what technologies will they take with them, further eroding U.S. technological dominance?

Other ethical issues regarding the top-of-the-value-chain strategies should also be brought out. First, one might challenge the ethics of trying to build a technologically dominant United States. (And, of course, efforts to achieve it may isolate the United States from the global system and challenge other countries to seek similar "home-country-first" goals). Beyond that, we believe U.S. attempts to dominate top-of-the-value-chain technologies will weaken the global ability to create technologies beneficial for all human kind. Surely medical breakthroughs available to the world are beneficial to all people, including Americans, wherever the breakthroughs take place. For that matter, would most Americans have been better off if the U.S. had somehow prevented the movement of Sony, Samsung, and other firms into the production of innovative consumer electronics products? Or would the U.S. automakers have improved quality and reduced fuel consumption without the competition from Japanese manufacturers? (Some of which now have more U.S.-made content than cars built by U.S. companies.)[12]

---

[12]The Ford Escape, for example, requires 2,500 U.S. workers and incorporates 65 % U.S.-manufactured parts as compared to more than 6,000 U.S. workers required to build a Toyota Camry, which contains 80 % U.S.-manufactured parts. For every 100 cars sold in the U.S., the Ford Escape

We believe that both ethics and enlightened self-interest argue for the U.S. (and the other currently richest countries) to develop a framework for achieving economic growth and prosperity based on "collaborative advantage" (Lynn and Salzman 2006). In brief, this is an approach that builds strength through participating in the global supply of human capital and innovation in collaboration with other nations. Rather than taking a zero-sum approach to innovation, economic growth, and prosperity, this approach is based on mutual-gain strategies in which the growth in global markets provides expanding economic and job opportunities in all countries. The goal for U.S. policy makers would be to benefit U.S. citizens through giving them a smaller piece of a much bigger pie, much as free trade policies historically have been far more beneficial to most people than mercantilistic policies. As such these policies would emphasize different approaches to S&T education within the United States, and would promote efforts by the United States to help build a "global commons" for the creation of beneficial technology.[13] In doing so it would also minimize some of the ethical dilemmas and value conflicts confronting many S&T people regarding nationalism and firm interest.

The United States is currently the best positioned country, we would argue, to lead this effort to establish a "global commons" of mutually beneficial global innovation and S&T workforces because of its history of openness, diversity, and free flow of knowledge, and because for the time being it is home to companies that are now leaders in developing globally distributed innovation systems (Lynn and Salzman 2006). Moreover, the United States still has knowledge and capacities within its universities and organizations that are not available in the emerging economies. As U.S. universities globalize, they can provide the highest-level knowledge, personnel, and experience to emerging economy universities and foster a tradition of the academic commons of science and collaboration.

2.6 Policies to Meet New Cognitive and Skill Requirements

Over at least the past 10–15 years, organizational, technological, and business strategy changes have been requiring new skills for engineers and other technical workers. The de-integration of technology development requires engineers to work across organizational boundaries with suppliers. Products that incorporate or have tightly integrated technology of different types, such as electronic and mechanical technology, or different materials, require engineers to work across disciplines, both within and outside of engineering. Business strategy that places more emphasis on

---

provides 12 U.S. jobs as compared to 20 U.S. jobs for every 100 Toyota Camrys sold in the U.S. http://www.theautochannel.com/news/2008/07/27/094534.html

[13]Nobel laureate Elinor Ostrom (1990) gives a variety of case studies set in different contexts and lays out a theoretical framework showing how institutions evolve into structures promoting collective action.

market-driven technology development also requires engineers to understand the business drivers as well as the technical drivers of product or service development.

These different boundary-spanning skills and abilities are increasingly important, especially when firms are systems integrators or are at the higher value-added part of the development chain, but also when they are playing supporting roles in technology development. Managers in our interviews typically said that technical skills were fairly easy to find and were not a distinguishing criterion between candidates. Setting good employees apart were their ability to communicate their ideas, to work with others on a team and with non-engineers, and other related social skills. These skills reflect the changes in the nature of engineering work, ranging from greater teamwork, working across disciplines, and interacting with customers and suppliers in developing and acquiring technology. An engineering manager we interviewed at a large electronics MNE said the best hire he had made in recent years was of a young engineer who was distinguished not by his technical abilities, but by his enthusiasm to travel and work with people from other cultures (Lynn and Salzman 2002).

Our research suggests that the skills S&T job applicants and workers lack are communication skills that enable employees to work across organizational and, increasingly, national boundaries, coordinate and integrate technical activities, and navigate the multidisciplinary nature of today's technical work. While a solid math, science, and technology education is necessary to form the foundation for skills required by S&T workers, globally competitive education must go far beyond training technically competent graduates. A broad education that incorporates a range of technical and social science and humanities knowledge is important for developing a globally competitive workforce (e.g., see Hill 2007). In this, the United States currently has an advantage over the emerging economies.

In summary, we consistently find employers in technology firms most valuing the boundary-spanning skills that require adroit communications skills and an ease at working outside of a narrow field of expertise or technical training. In nearly all cases managers found a plentiful supply of technically qualified applicants and hiring decisions were made on the basis of their non-technical skills. While many of these skills can be provided through broad-based, multi-disciplinary education, some of the skills appear to come from cross-national experiences. In most cases, although these people were educated in the United States they were not born here and had lived in more than one culture. Perhaps these skills can be taught, but they may also require educators to incorporate cross-national experiences as part of technical training. Such abilities are likely to foster an appreciation for other cultures, and lead to more ethical treatment of those who are different.

2.7 Conclusions

The technological challenges now facing the United States are immense, but they are different from those that faced the country in the past. Today's challenges will require new strategies for national well-being that do not depend on maintaining global dominance in science, technology, and innovation. The U.S. did enjoy such dominance for half a century and more after World War II, not least because previous technology champions like Germany, England, and France had had their innovation systems crippled by war. Although the depth and breadth of U.S. science, engineering, and innovation continue to be unmatched, the globalization of technology and innovation work by firms and the increasing globalization of U.S. universities are leading to the rise of new centers of innovation across the globe. The global monopoly of innovation by the U.S., EU, and Japan has ended.

Many influential policy advocates are alarmed by the growing technological strength of the emerging economies. Implicit in their proposals is a strategy of having the United States pursue nationalist, zero-sum technology policies with the intent of disadvantaging other nations, particularly the emerging economies. The outcome of these policies would be to entice more young Americans to seek S&T education and training while at the same time undermining the career prospects in those jobs. Moreover, these policies would lead to the U.S. absorbing and sequestering top talent from other countries.

Both strategies raise ethical and practical issues. While improving the science and math education of Americans from kindergarten through high school is surely a worthy goal, there is a problem if the consequence is to divert attention and resources away from other worthy educational goals, such as teaching students to cope with the increasing diversity of the people they will be interacting with both in their personal lives and in their jobs. There is also an ethical and practical problem if the goal of producing a large number of S&T workers who are globally competitive means continuing neglect of the much greater numbers needed to effectively *implement* innovation. Aside from that is the problem of using of policy interventions to divert students pursuing other careers into S&T fields and away from more rewarding alternatives? And, while the U.S. has a rich history of providing a place to live and work for foreigners pushed out of their homelands for political reasons or attracted to the U.S. to achieve fulfillment of their potential as S&T workers, it is ethically questionable for U.S. policies to actively promote brain drains, taking the best and brightest from emerging economies who have invested heavily in them and need them to build prosperity at home. Further, as we have shown, by bringing in large numbers of entry-level S&T workers, some visa policies would result in lower salaries and thus undermine the job quality and strength of the domestic U.S. S&T workforce.[14] At the same time, it would promote the outward flow of technologies financed by U.S. taxpayers.

---

[14]First of all, it might be noted that there is no evidence that the United States, whatever the faults of its education system, is not already producing enough talented people to fill university S&T programs. Secondly, even if these measures would increase the *supply* of potential American S&T workers, they would not address the lack of *demand* for such workers. Large numbers of American graduates from S&T programs do not seek employment in S&T jobs, and many of those who do become S&T workers are quick to move to jobs outside S&T. Often non-S&T jobs are more attractive because they pay more. Ignoring the second part of the equation, where firms are locating their jobs, can only further undermine the strength of the U.S. S&T workforce.

But putting aside any ethical qualms, what are the chances that these proposed policies would actually work? If we increase the supply of S&T workers in the U.S. what happens if there is no increase in demand? Perhaps wages would go down, and this might increase the number of jobs available as U.S. MNEs seeking low cost S&T workers return jobs to the U.S. But it would also increase frustration among new graduates, as is already happening with holders of graduate degrees who feel trapped in endless unrewarding postdoctoral programs (Freeman et al. 2004; Freeman 2006; Stephans 2012). Perhaps MNEs could try to make S&T jobs more attractive by raising salaries, but as we saw in our discussion of Apple, this would raise costs and make doing S&T work in the U.S. less attractive for corporate decision makers and investors. Perhaps we could have policies to compel U.S. MNEs to locate more jobs in the U.S., but how would this be done so as not provoke retaliation? And might not some "U.S. multinationals" respond by further dissolving their ties to the U.S. (just as some have moved their registered headquarters offshore in pursuit of more favorable tax policies)? Can we assume that China, India, and other countries will stand idly by as the U.S. seeks to entice and retain some of their most talented S&T workers? And can we assume that large numbers of the talented S&T workers from the emerging economies will not return home when opportunities present themselves? And why shouldn't they?

Our analysis suggests several education and policy recommendations that will strengthen U.S. science, technology, and innovation, while maintaining the ethical high ground. This list includes several current initiatives that we believe would help strengthen the U.S. workforce.

1. **Emphasize a broad education rather than a narrow technical education**. Math and science skills are not what employers report as being in short supply among their professional and technical workforce. An overemphasis on math and science could lead to the exclusion of the skills employers report most needing among their S&T workers. At the same time, it is important to broaden the content and improve the pedagogy of science and math throughout the education system, at primary, secondary, and college levels. There are a number of efforts under way to improve science, math, and engineering education; additional support and diffusion of new curricula would be beneficial, as would improving other educational areas. It is important to distinguish between improving the quality for all students from policies focused on increasing the quantity and/or focusing on the upper levels of the student population as a means of increasing the S&T workforce. The globalization of engineering activities will also pose other ethical issues for engineers and engineering managers – the fair treatment of women and minorities in cultures accustomed to even higher levels of inequality than our own. It is essential that we educate our engineers and managers to develop a greater sensitivity to these issues and a greater ability to cope with them.

2. **Encourage complements rather than substitutes in the labor market through immigration policy.** The focus on absorbing more foreign workers and students, either through targeted work-based immigration policies or for

graduates of specific disciplines, has little justification in terms of domestic labor market needs. In fact, as discussed above, it directly undermines the position of those in the domestic S&T labor market (e.g., see Salzman et al. 2013). At the same time, decades of brain drain and investment in higher education have provided the U.S. some of the strongest universities in the world. Policies for collaborative advantage would support the globalization of student flows with the strategic and ethical objective of developing global linkages and liberalizing educational and S&T policies in universities in other, often restrictive countries. Foreign graduates who return to their home or other countries strengthen U.S. ties and the global S&T commons if those graduates develop practices and policies that facilitate the global flow of innovation. Their U.S. education can provide that foundation, and thus true global flows of S&T workers, rather than a U.S. sequestration of graduates.

3. **Establish international labs, similar to the model of the U.S. national labs.** Taking the lead in developing the structure and terms of participation in the global commons will provide the United States continued access to innovation and knowledge around the globe. It will also create new and exciting opportunities for U.S. S&T workers as well as integrate global S&T workers into networks in which the United States participates. This is one means of benefiting from global human capital development without substituting it for domestic S&T workers. This model reflects the current strategies of many U.S. firms and universities and the current structure of globalization in which knowledge and innovation flow across national boundaries.

4. **Focus innovation and technology policies on pressing global problems and technology that meet global needs.** Understanding the dual innovation frontiers—not just high-end technology—and addressing global problems should be a key aspect of R&D policy. In particular, a focus on innovation under resource constraint, such as the resource constraint of using hydrocarbon energy sources and their environmental impacts, will lead to innovations applicable to emerging markets. Many firms are doing this, but in other countries. Developing leading expertise in the U.S. will keep the United States engaged in global technology development.

5. **Develop policy frameworks based on collaborative advantage and participation in the global commons of innovation.** Trying to achieve "dominance" or "supremacy," will not garner the support of other countries or of foreign U.S. S&T workers who would like to contribute to the development of their countries of origin. The above proposals would promote U.S. national interests through mutual-gain policies giving the U.S. a collaborative advantage. Above all, we believe a change in rhetoric and policy focus will then lead to the development of more specific initiatives. The current focus on U.S. techno-nationalist policies leads us in the wrong direction and it alienates many of the foreign nationals on whom our companies and universities depend, and who indeed are a vital part of the U.S. innovation system.

References

American Electronics Association. (2005). *Losing the competitive advantage? The challenge for science and technology in the United States.* Washington, DC: American Electronics Association.

Associated Press. (2005). Sun Microsystems expands around globe. CNBC news. http://www.nbcnews.com/id/7791982/ns/technology_and_science-tech_and_gadgets/t/sun-microsystems-expands-around-globe/. Accessed 10 Feb 2014.

Business Week. (2007). Where are all the workers? *Bloomberg Business Week magazine.* http://www.businessweek.com/stories/2007-04-08/where-are-all-the-workers. Accessed 10 Feb 2014.

China Mobile. Investor Relations. (2014). http://www.chinamobileltd.com/en/ir/operation.php?section=number&year=2014. Accessed 18 Feb 2015.

Committee on Prospering in the Global Economy of the 21st Century: An Agenda for American Science and Technology, National Academy of Sciences, National Academy of Engineering and Institute of Medicine. (2007). *Rising above the gathering storm: Energizing and employing America for a brighter economic future.* National Academy of Sciences, Washington, DC http://www.nap.edu/catalog/11463/rising-above-the-gathering-storm-energizing-and-employing-america-for

Cusumano, M., & Selby, R. (1995). *Microsoft secrets: How the world's most powerful software company creates technology, shapes markets and manages people.* New York: Touchstone.

Davis, G. (1997). *Mathematicians and the market.* Hanover: Dartmouth College.

Espenshade, T. J. (1999). *High-end immigrants and the shortage of skilled labor.* Working Paper No. 99-5. Princeton: Office of Population Research, Princeton University.

Fierce Wireless. Grading the Top 8 Wireless Carriers in the Third Quarter of 2014. http://www.fiercewireless.com/special-reports/grading-top-8-us-wireless-carriers-third-quarter-2014?confirmation=123. Accessed 18 Feb 2015.

Freeman, R. B. (1976). *The overeducated American.* New York: Academic.

Freeman, R. B., Emily Jin, & Chia-Yu Shen. (2004). *Where do new U.S.-trained science-engineering PhDs come from?* NBER Working Paper No. 10554. Cambridge, MA: National Bureau of Economic Research.

Freeman, R. B. (2006). Does globalization of the scientific/engineering workforce threaten U.S. economic leadership?". In *Innovation policy and the economy* (Vol. 6). Cambridge, MA: NBER/MIT Press.

Gallagher, J. (2011). Emerson's David Farr is citizen of the year. St. Louis post-dispatch. Feb 13. http://www.stltoday.com/business/local/emerson-s-david-farr-is-citizen-of-the-year/article_d6948a59-5707-51ad-b91a-872ed2f7fae4.html.

GSMA Intelligence. Global Mobile Phone Data. https://gsmaintelligence.com. Accessed 18 Feb 2015.

Hill, C. T. (2007). The post-scientific society. *Issues in Science and Technology*, Fall, 78–84.

Iritani, E. (2002). High-paid jobs latest U.S. export: Firms' shifting of technical work to Mexico and China to cut costs bodes ill for many laid-off Americans. *Los Angeles Times.* Apr 2. http://articles.latimes.com/2002/apr/02/news/mn-35803.

Lowell, B. L., & Salzman, H. (2007). *Into the eye of the storm: Assessing the evidence on science and engineering education, quality, and workforce demand.* Research Report for The Urban Institute. http://www.urban.org/research/publication/eye-storm

Lynn, L., & Salzman, H. (2002). What makes a good engineer? U.S., German, and Japanese perspectives. Proceedings of European applied business research conference, Rothenberg, June.

Lynn, L., & Salzman, H. (2004). Third generation globalization: The new international distribution of knowledge work. *International Journal of Knowledge, Culture and Change Management, 4,* 1511–1521.

Lynn, L., & Salzman, H. (2006). Collaborative advantage. *Issues in Science and Technology,* Winter, 74–82.

Lynn, L., & Salzman, H. (2007). The real technology challenge. *Change: Magazine of Higher Learning*. July/August.

Lynn, L., Meil, P., & Salzman, H. (2012). Reshaping global journal technology development: Innovation and entrepreneurship in China and India. *Journal of Asian Business Studies, 6,* 143–159.

McKinsey Global Institute. (2013). *Urban world: The shifting global business landscape*. Report, October.

National Center on Education and the Economy. (2007). *Tough choices or tough times: The report of the new commission on the skills of the American workforce*. Washington, DC: National Center on Education and the Economy.

National Science Foundation. (2006). *Science and engineering indicators 2006*. Appendix table 2–6. http://www.nsf.gov/statistics/seind06/pdf_v2.htm. Arlington, VA: National Science Foundation.

Ostrom, E. (1990). *Governing the commons: The evolution of institution for collective action*. New York: Cambridge University Press.

Perry, T. (2014). After the sun (Microsystems) sets, the real stories come out. *IEEE Spectrum*. May 30. http://spectrum.ieee.org/view-from-the-valley/at-work/tech-careers/after-the-sun-microsystems-sets-the-real-stories-come-out

Perry, T. (2015). Massive worldwide layoff under way at IBM. *IEEE Spectrum*. Feb 3. http://spectrum.ieee.org/view-from-the-valley/at-work/tech-careers/massive-worldwide-layoff-underway-at-ibm

Porter, M. (1998). Clusters and the new economics of competition. *Harvard Business Review*, November–December: 77–90.

Salzman, H. (2000). *The information technology industries and workforces: Work organization and human resource issues*. Committee on Workforce Needs in Information Technology.

Salzman, H. (2013). What shortages? The real evidence about the STEM workforce. *Issues in Science and Technology* (National Academy of Science policy magazine), Summer 2013: 58–67

Salzman, H., & Lowell, L. (2008). Making the grade. *Nature, 453*, 28–30. http://www.nature.com/nature/journal/v453/n7191/pdf/453028a.pdf.

Salzman, H., Kuehn D., & Lowell, B. L. (2013). Guestworkers in the high-skill U.S. labor market: An analysis of supply, employment, and wage trends. Economic Policy Institute. http://www.epi.org/publication/bp359-guestworkers-high-skill-labor-market-analysis/. Accessed 10 Feb 2014.

Saxenian, A. (2000). The origins and dynamics of production networks in Silicon Valley. In M. Kenney (Ed.), *Understanding Silicon Valley*. Stanford: Stanford University Press.

Stephan, P. (2012). *How economics shapes science*. Cambridge, MA: Harvard University Press.

Teitelbaum, M. S. (2008). Structural disequilibria in biomedical research. *Science, 321*(5889), 644–645.

Teitelbaum, M. S. (2014). *Falling behind? Boom, bust, and the global race for scientific talent*. Princeton: Princeton University Press.

Von Hippel, E. (1994). 'Sticky information' and the locus of problem solving: Implications for innovation. *Management Science, 40*, 429–439.

Wall Street Journal. (2006). High-tech titans unite on lifting visa caps. *Wall Street Journal*, 14 June, p. A2.

Wiggin, A. (2010). Atlas isn't shrugging, he's outsourcing. *Forbes*. May 18. http://www.forbes.com/sites/greatspeculations/2010/05/18/atlas-isnt-shrugging-hes-outsourcing/

Wikipedia. List of countries by number of mobile phones in use. http://en.wikipedia.org/wiki/List_of_countries_by_number_of_mobile_phones_in_use. Accessed 18 Feb 2015.

Chapter 3
International Ethics and Failures: Case Studies

Norbert Delatte

Abstract Many engineering codes of ethics, such as that of the American Society of Civil Engineers, require that engineers "shall hold paramount the safety, health and welfare of the public." However, structures such as buildings and dams can fail with considerable loss of life. If engineers are held responsible for these failures, penalties may be imposed by criminal or civil courts, by licensing boards, or by professional societies. In this paper five cases are reviewed – the Hyatt Regency Walkway Collapse (1981), United States, the Malpasset Dam (1959), France, the Vaiont Dam Landslide (1963), Italy, the Sampoong Superstore (1995), Republic of Korea, and the Rana Plaza Building (2013), Bangladesh. Following the collapse of the Hyatt Regency Walkways, which killed 114, the engineers were not charged with a crime, but were stripped of licenses by state licensing boards and one was expelled from professional society membership. The failure of the Malpasset dam, which killed 421 people in the subsequent flood, resulted in both criminal and civil court actions. The landslide in the Vaiont Dam reservoir and subsequent flooding resulted in over 2,000 deaths. Eleven engineers and others were criminally charged, three were found guilty, and two served short jail terms. Following the collapse of the Sampoong Superstore which killed almost 500, executives of the firm that owned the store served 7 and 10 ½ year jail sentences. In addition, 12 local building officials were found guilty of taking bribes. Recently, another tragic building collapse has occurred in Bangladesh. These cases contrast different approaches taken to toward public safety in different countries.

Keywords Structural collapse • Dam failure • Forensic engineering • Engineering ethics • Corruption

N. Delatte (✉)
Cleveland State University, Cleveland, OH 44115, USA
e-mail: n.delatte@csuohio.edu

© Springer International Publishing Switzerland 2015
C. Murphy et al. (eds.), *Engineering Ethics for a Globalized World*, Philosophy of Engineering and Technology 22, DOI 10.1007/978-3-319-18260-5_3

3.1 Introduction

The work of engineers has vast potential to benefit society, but has the potential to injure and kill as well. Around the world, engineers in general, and civil engineers in particular, hold a responsibility to protect the safety, health, and welfare of society. Many of a civil engineer's works, such as water purification systems, have direct positive impacts on health. Other, such as buildings and transportation systems, benefit the welfare of society by improving comfort and commerce.

That said, the work of engineers also carries considerable risk. Construction remains a dangerous profession, with many workers injured and killed every year. The general public is also at risk from engineering failures. Although thankfully rare, building and dam collapses have the potential to kill and injure with little or no warning.

This special burden of engineers is addressed in different ways around the world. In some countries, such as the U.S., engineers are expected to adhere to a code of ethics. In other countries, failures may be disciplined through the legal system. This chapter reviews one code of ethics, that adopted by the American Society of Civil Engineers (ASCE), and then reviews five case studies of building collapses and dam failures that led to loss of life, and their repercussions.

3.2 The ASCE Code of Ethics

In the United States, the conduct of engineers is overseen by professional societies, such as ASCE, and by licensing jurisdictions, which are generally states or territories. Each licensing jurisdiction has a Code of Ethics, although there are many similarities. Professional societies only have jurisdiction over their members, and the maximum penalty is suspension or expulsion from the Society. ASCE has many international members, who are potentially subject to disciplinary actions even if they don't live or work in the U.S. Licensing jurisdictions can discipline anyone who offers engineering services, regardless of whether the individual holds a professional engineer license. Penalties can include fines and revocation of licensure, or orders to cease advertising engineering services.

Many engineering codes of ethics, such as that of the ASCE, directly address the requirement of engineers to promote and preserve public safety. This Code of Ethics has seven Fundamental Canons. Fundamental Canon 1 of the ASCE Code of Ethics states that "Engineers shall hold paramount the safety, health and welfare of the public and shall strive to comply with the principles of sustainable development in the performance of their professional duties" (ASCE 2013). The word "paramount" suggests that this Canon supersedes all others, in case of conflict between different parts of the Code.

The "Guidelines to Practice Under the Fundamental Canons of Ethics" provide more detail. It further states that "Engineers shall recognize that the lives, safety, health and welfare of the general public are dependent upon engineering judgments, decisions and practices incorporated into structures, machines, products, processes and devices" (ASCE 2013).

Of the five Guidelines under Canon 1, (b) is particularly relevant to the protection of the public. It states that "Engineers whose professional judgment is overruled under circumstances where the safety, health and welfare of the public are endangered, or the principles of sustainable development ignored, shall inform their clients or employers of the possible consequences" (ASCE 2013). This requires that engineers take an active role to inform their clients or employers in cases of danger.

It is, unfortunately, not specific as to what the engineer should do if the client or employer does not act on the information to protect the safety, health and welfare of the public. It is the author's opinion that, in that case, the engineer should take measures to inform the authorities and the public.

The National Society of Professional Engineers (NSPE) also publishes a code of ethics, which is very similar to the ASCE Code. NSPE Fundamental Canon 1 states the engineers shall "Hold paramount the safety, health, and welfare of the public." The first Rule of Practice under that Canon requires that "If engineers' judgment is overruled under circumstances that endanger life or property, they shall notify their employer or client and such other authority as may be appropriate" (NSPE 2013). The NSPE addition of "such other authority as may be appropriate" seems like a good addition.

3.3 Hyatt Regency Walkway Collapse (1981), United States

For engineers in the United States, one of the defining collapses for the profession occurred on July 17, 1981. The atrium of the Kansas City Hyatt Regency was crossed by three suspended walkways at the 2nd, 3rd, and 4th floors. The 4th floor walkway was over the 2nd floor walkway, and the 3rd floor walkway was offset from the other two.

During a Friday afternoon tea dance, the 4th floor suspended walkway fell 114 were killed, and nearly were 200 injured. Figure 3.1 shows the two collapsed walkways. Pfatteicher (2000) provides a dramatic account of the collapse.

What had happened was immediately apparent. The 4th floor walkway had pulled loose from its hanger rod, and fallen onto the 2nd floor walkway, and then both fell to the floor. The connection was clearly deficient. Why it had been built that way turned out to be a bit more complicated.

3.3.1 The Story Usually Told

This has been used as a cautionary tale by engineering faculty for nearly two decades. Figure 3.2 shows the original "design" and the actual construction. The original design could not be built, because there really is no good way to get the

Fig. 3.1 Hyatt regency walkways following collapse (Wikipedia Commons 2013)

supporting nut into the middle of the hanger rod. Because the original design could not be built, the detailer requested a change.

This change doubled the force transferred at the beam-to-nut-to-rod connection, shown as "2P on nut" in Fig. 3.2. This change in the force is evident from the free body diagram of the connection. The engineer was contacted on three separate occasions with concerns about the connection, and provided reassurances that it was safe. The engineers concerned, Jack Gillum and Daniel Duncan, were not prosecuted but had their Missouri Professional Engineer (PE) licenses revoked. Several alternate designs could have solved the problem.

3.3.2   An Alternative View

The story, as usually told, is a bit misleading. There is one point in the case study that is often missed. The "as-built" or "actual construction" view was never shown on any project documents, and was only drawn after the collapse (Delatte 2009). It is therefore incorrect to refer to it as an "original design," since in fact there was never a design of this connection.

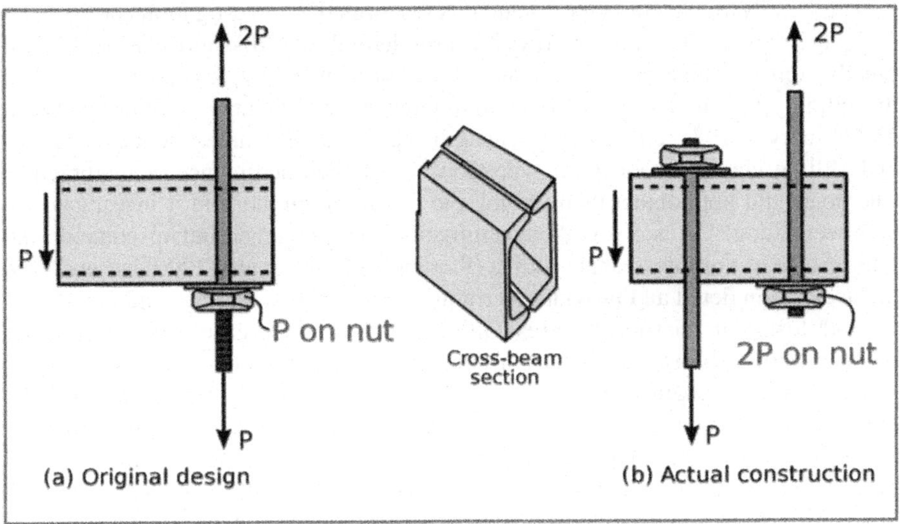

Fig. 3.2 Comparison of original design and as-built connection (Wikipedia Commons 2013)

Luth (2000) provides a detailed discussion of the sequence of events that led to the failure. It was extraordinarily convoluted, even for a project of this size. His account, which is also summarized by Delatte (2009), establishes that the design of this critical detail became lost in a chain of miscommunications with a troubled project.

In essence, this connection, like most other connections in the building, was passed on to the connection detailer for design. Because some critical information disappeared when the drawings were copied, at some point the detailer assumed that the connection had already been designed (Luth 2000; Gillum 2000).

3.3.3   Board and ASCE Actions

The ASCE Code of Ethics, in pretty much its present form, was adopted in 1975 (ASCE 2013) and was thus very new when this incident occurred. The major change had been the addition of Fundamental Canon 1, discussed above. This case provided the first test of the new ethics code and licensing law interpretations (Pfatteicher 2000).

The public in Kansas City demanded justice for the victims – nearly all of the dead were locals. An attempt was made at criminal prosecution. In December 1983, the county prosecutor and U.S. Attorney found insufficient evidence to convict anyone of criminal negligence, and no charges were filed. Insurance companies eventually paid more than $100 million to settle civil lawsuits (Luth 2000).

The U.S. National Bureau of Standards performed an investigation and issued a 377 page report. Once the report had been released, the Missouri licensing board quietly convened a hearing. The policy of the board over 40 years had been to avoid disciplinary action. The board filed complaints against Duncan, Gillum, and firm on February 4, 1984, and then presented the case to Judge James Deutsch. Duncan and Gillum claimed poor communication, rather than negligence or malpractice. The board did not contact them during the investigation, nor did it investigate the architects. Judge Deutsch found both guilty of gross negligence and misconduct, and Gillum also of unprofessional conduct (Pfatteicher 2000). Luth (2000) has analyzed the decision in detail and provides a critique of the Judge's findings and logic.

Both lost their Missouri licenses. Gillum lost 24 of 28 licenses in other jurisdictions as well. California granted reinstatement of Gillum's license in 1994, and he began to practice engineering again. Duncan has not practiced engineering since the board revoked his license. Some engineers have severely criticized Judge Deutsch's findings, claiming that they oversimplified the complexity of the modern fast paced project design and construction environment (Pfatteicher 2000). On the other hand, others felt that the punishment was more than justified (Rubin and Banick 1987).

The ASCE Committee on Professional Conduct held a confidential hearing in summer 1986. The Committee voted unanimously to expel Gillum; Duncan was not an ASCE member and thus could not be disciplined by ASCE. The ASCE Board of direction suspended Gillum for 3 years, with the details kept confidential. Gillum has not applied to rejoin ASCE (Pfatteicher 2000).

Although it must have been painful, Gillum subsequently made a number of presentations for seminars and groups of engineering students about this case as a cautionary tale. In 2000, he presented a paper at the 2nd ASCE Forensic Engineering Congress (Gillum 2000).

Both Gillum and Luth believe that although some lessons were learned from the incident, there were many other important lessons that were not learned. Luth (2000) observes that structural safety is much too important to use a scope of services defined by a low bid process, that city building departments do not and cannot provide enough checking on major projects, and that good structural engineering practice should not be defined after the fact by the legal profession on a case-by-case basis.

3.4   Malpasset Dam (1959), France

The Malpasset Dam was a 66 m high concrete arch dam in Southern France, across the Reyran River. It was located approximately 7 km north of Fréjus on the French Riviera. Like many dams, it was built for multiple purposes, in this case for irrigation, drinking water, and flood control. The dam was built between April 1952 and December 1954. The Malpasset Dam was a continuation of the work of the famed designer André Coyne, who had slowly been raising the allowable compressive stresses within arch dams. At Malpasset, the allowable compressive

Fig. 3.3 The failed Malpasset Dam (Wikipedia Commons 2013)

stress was increased from 2.4 MPa to 4.7 MPa. Arch dams require much less concrete than traditional gravity dams, because instead of using the heavy mass of the concrete to hold back water, they use a thin efficient arch form. When it was completed, Malpasset was the thinnest arch dam in the world (Ross 1984, p. 127).

A series of rainstorms filled the reservoir quickly in late fall of 1959. At 9:14 p.m. on December 2, 1959, the Malpasset dam failed explosively, giving rise to a flooding wave more than 40 m high. Only a little portion of the arch of the dam still remained in its original position. Four hundred twenty one deaths were attributed to the failure. The failed dam is shown in Fig. 3.3.

The technical cause of the failure was revealed during the investigations. A thin clay filled seam in the rock behind the left abutment allowed the abutment to shift. This displacement and the loss of support led to the cracking at the center of the dam (Levy and Salvadori 1992, pp. 166–172). The clay seam was only 25–50 mm wide, and had not been found during the preconstruction foundation investigations. The prosecutor, after the failure, alleged that this seam should have been taken into account in the dam's design.

Post failure evaluations revealed that the dam was founded on a large block of rock that slid out of the left abutment on an existing geologic fault plane. When the foundation slid as a result of the loads from the dam and reservoir, the left side of the dam went with it.

The 69-year-old engineer of the Malpasset Dam, André Coyne, died 6 months after the disaster. The engineer who accepted the dam on behalf of the Agriculture Department, Jacques Dargeou, was indicted for negligent homicide.

Dargeou demanded a new investigation. The new investigation faulted Dargeou for not exploring the dam's foundation better but also implicated the late Coyne. After a long trial, Dargeou was acquitted. After this event, a group of 240 relatives of victims sued four of the engineers involved.

Interesting new facts emerged. A month before the collapse, changes in the shape of the dam had been documented in a photo survey. The engineer who had taken over from Dargeou did not consider this information significant but had been slowly starting an investigation. In retrospect, it is likely that this was a very clear indication of impending dam failure. It is also interesting that the French Ministry of Agriculture concluded that this incident should in no way diminish confidence in the safety of thin arch dams (Ministere de l'Agriculture 1960).

3.5 Vaiont Dam Landslide (1963), Italy

Following the 2nd World War, Italian engineers built an extensive system of dams, reservoirs, and hydroelectric powerhouses in the Piave River Valley, high in Italian Alps. Part of this work was the Vaiont Dam, another thin concrete arch dam much like the Malpasset Dam. The dam was located on the Vaiont River under Monte Toc, 100 km north of Venice, Italy. The original idea was conceived in 1920 and was authorized on October 15th 1943. Construction began in 1957 and was completed in 1959. The filling of the basin began in February 1960, despite studies by experts saying that the entire side of Mt. Toc was unstable and would likely collapse into the basin if filled completely. The Vaiont dam is one of the tallest dams in the world, standing 262 m high.

Ironically, the Vaiont Dam was completed at about the same time that the Malpasset Dam failed. Although the Vaiont Dam incident was also a geotechnical failure, it was very different from Malpasset. A slope stability failure is more commonly known as a landslide, particularly among non-engineers. This type of failure occurs when the weight of a soil mass overcomes the soil's shear resistance along a failure plane. Water within soil increases its unit weight, while reducing the shear strength. As a result, water and water pressures often play a role in triggering a slope stability failure. The Vaiont Dam disaster of 1963 was a classic slope stability failure.

The Vaiont Dam held back 169 million cubic meters of water. In June 1957, the owner increased the capacity of the reservoir by 30 %, anticipating future needs for more electrical power. On March 22, 1959, a landslide of three million cubic meters of rock at nearby Pontesei Reservoir killed one person. This led to concern about the stability of the sides of the Vaiont reservoir.

An investigation by the young geologist Edoardo Semenza found evidence of an ancient landslide, along an uncemented mylonitic zone. He also identified some other important geologic features, such as a 1.5 km zone of uncemented cataclasistes along the base of the left wall of the valley, along with solution cavities, sinkholes, and springs. He also found that ancient landslide masses had filled the valley, and then had been cut into two by the new Vaiont stream, and, further, that the southern slope of Mt. Toc had a "chair like" structure of bedding planes, dipping steeply at the top and more shallowly near the base, and that a fault separated the in situ rock mass from the ancient landslide (Genevois and Ghirotti 2005). In Semenza's opinion, these geologic features and the evidence of a landslide in the past made a future landslide likely. However, the dam's designers thought that a landslide was unlikely, based on opinions from other experts. They also discounted Semenza's opinions because of his youth.

In 1960, the owner began filling the dam and monitoring the earth movements. The movements increased as the dam filled. On November 4, 1960, a landslide deposited 750,000 cubic meters of soil into the reservoir. Clearly, with this incident and the earlier one at the Pontesei reservoir, the potential for a landslide had been demonstrated (Delatte 2009).

The dam owner thought that the movement could be controlled by lowering the water. It was observed that the earth movements increased whenever the reservoir level rose, and decreased then the level dropped. In addition, boreholes were drilled to reduce water pressure. It was believed that relieving the water pressure would reduce the risk of a landslide.

After nearly 3 years of sporadic, slow slope movements that began with the first filling of the reservoir, a catastrophic landslide suddenly occurred on the southern slope of Mt. Toc on the 9th October 1963 at 10:39 pm local time. This occurred during the third reservoir emptying operation. The whole mass collapsed into the reservoir in less than 45 s. The landslide moved down the mountain at a speed of up to 25 m/s.

As the landslide filled the reservoir, a wall of water went over the dam, sending a 70 m wall of water down Vaiont gorge. It destroyed the town of Longarone downstream, and severely damaged or destroyed the hamlets and villages of Villanova, Codissago, Pirago, and Fraseyn. Two thousand and forty three people were killed, including 58 of the utility's employees (Wearne 2000, pp. 213–214).

The population of Longarone before the disaster had been about 4,600. The flood also knocked out many access routes, hampering rescue operations (Ross 1984, p. 132). Figure 3.4 shows the dam, looking up from Langarone. Ironically, the dam itself suffered almost no damage, with only a small bit of concrete removed at the left abutment. Figure 3.5 shows the reservoir filled in behind the dam.

3.5.1 Italian Government Committee of Inquiry

The Italian government commissioned a committee of inquiry, which blamed "bureaucratic inefficiency, muddled withholding of alarming information, and buck-passing among top officials." The prime minister of Italy suspended a number of public officials, including the province chiefs and civil engineers of Belluno and Udine.

Fig. 3.4 The Vaiont Dam, as seen below from the town of Langarone (Wikipedia Commons 2013)

More than 4 years after the disaster, the public prosecutor of the province of Belluno charged 11 men with crimes ranging from manslaughter to negligence. Proceedings focused more on assigning blame than on determining the technical cause of the failure.

By that time, two of the men charged had already died. A third committed suicide the day before the trial was to begin. Charges included ignoring consultants' cautions and failing to fully investigate the prior earlier earth movements in the slide area. The prosecutor asserted that each of the men charged could have controlled the situation and prevented the disaster. Three of the 11 charged were found guilty, and two served short jail terms (Delatte 2009).

The landslide was investigated by Hendron and Patton (1985) of the U.S. Army Corps of Engineers Waterways Experiment Station. They found a thin layer of weak clay, which provided a lubricant for the landslide. Higher water pressures behind the slide plane correlated with higher rainfall and higher reservoir levels. Sinkholes

Fig. 3.5 Behind the Vaiont Dam – reservoir filled in by landslide (Wikipedia Commons 2013)

in the karstic plain of Mt. Toc allowed water to infiltrate. Previous boreholes for testing did not go deep enough. Semenza's observation of an ancient landslide was confirmed by later investigations.

Finally, Mt. Toc in the local dialect means crazy – the locals were aware of ground movements. Sometimes local knowledge is particularly important for engineering.

One important difference between the Malpasset and Vaiont Dam incidents is that, in the latter case, there seemed to be much clearer warnings of the impending landslide that could have and should have been heeded and used to evacuate the population from below the dam. Whereas the problems with the Vaiont Dam were documented and discussed in detail before the incident, the potential warning from the change in shape of the Malpasset Dam only became widely known after the failure.

3.6 Sampoong Superstore (1995), Republic of Korea

The collapse of the Sampoong Superstore in Seoul, South Korea, represents an example of a structural collapse attributed in large part to corruption. The Sampoong department store opened in December 1989. It was a nine-story building with four basement floors and five above grade. The building was laid out in two wings (north and south) connected by an atrium lobby. By the mid-1990s the store's sales amounted to more than half a million U.S. dollars a day (Wearne 2000, pp. 99–100).

Unfortunately, the store had been built on a landfill site that was poorly suited to such a large structure. Woosung Construction built the foundation and basement and then passed the project on to Sampoong's in-house contractors. Woosung had apparently resisted some proposed changes to the building plans, such as the addition of the fifth floor (Wearne 2000, p. 100).

Sampoong made significant changes to the structure. The most important was the conversion of the original use as an office block to that of a department store. Other changes included changing the upper floor from a roller-skating rink to a traditional Korean restaurant. Stricter standards had to be met for fire, air conditioning, and evacuation. Although the structure apparently met all building code requirements, the revised design was radically different from the original (Wearne 2000, p. 100). In order to install the escalators needed for the department store, parts of columns were cut away (The Dawn 2004).

The Korea Times noted "Indeed, it was the popularity of Sampoong's eateries that indirectly caused the disaster. According to an established tradition, no Korean department store is complete without an 'eating gallery.' The Sampoong eating gallery enjoyed remarkable success, so the owners decided to improve it. They arranged an artificial pond with a life-size watermill and installed a lot of heavy air-refrigeration equipment. The result was a huge weight overload that made the entire construction unstable. Renovations to the underground parking lot – conducted without any approval – further increased this instability, so by 1995 the fashionable Sampoong Emporium was a disaster waiting to happen" (The Dawn 2004).

The building was put into service. "For five and a half years business thrived. In June 1995 the store passed a regular safety inspection. But within days there were signs something was seriously wrong: cracks spidering up the walls in the restaurant area; water pouring through crevices in the ceiling. On June 29 structural engineers were called in to examine the building. They declared it unsafe. Company executives who met that afternoon decided otherwise. They ordered the cracks on the fifth floor to be filled and instructed employees to move merchandise to the basement storage area" (Wearne 2000, p. 100).

"Around 5 p.m., the ceilings of the fourth floor visually began to sink. The fourth floor was also closed, but trade did not stop until 5:50 p.m., when the entire building began to make cracking sounds. The alarms went off, and the evacuation began, but it was far too late" (The Dawn 2004). At 6:00 p.m. on June 29, the center of the building collapsed, similar to a controlled implosion, in about 10 s. The five-story north wing, about 91 m long, fell into the basement, leaving only the façade standing (Wearne 2000, pp. 100–102).

Customers were concentrated in the basement and in the fifth-floor restaurant. The customers and employees had no time to run. Some survivors were found in the wreckage, and one woman was brought out alive 17 days after the collapse. The overall death toll was 498 (Wearne 2000, pp. 100–107).

3.6.1 Investigation

Professors Chung and Choi, of Dan Kook University and Soongsil University, investigated the collapse, as reported by Wearne (2000, pp. 107–111). Their investigation found that the building was a flat slab structure without cross beams, and thus inherently nonredundant. The foundation was adequate. Large chunks of the structure had been cut away in the conversion from an office block. The spans between columns had been increased to almost 11 m for more sales space. The fifth floor restaurant added considerable weight, and consisted of 90 cm thick concrete. The restaurant floor plan not compatible with the lower floors – the columns did not line up, and as a result the load path to the foundation was compromised. Large, heavy water cooling blocks had been installed on the roof. To support the water coolers, the roof slab was thickened – but no columns were added or enlarged. Not long before the collapse, the water cooling blocks had been moved across the roof, causing cracks up to 25 mm wide.

There were other structural problems as well. The columns were only 62 cm wide versus the 89 cm in the design, and only 8 of 16 reinforcing bars were provided. The slab dead loads had been miscalculated – they were based on 10 cm thick floor slabs but the actual slabs were 3 or 4 times as thick. Some of the reinforcement had been left out, and the connections between slabs and floors were weak.

A technical paper on the collapse entitled "Lessons from the Sampoong Department Store Collapse" (Gardner et al. 2002) was published in the journal *Cement & Concrete Composites*. Their analysis verified that the collapse was due to the reduced depth and strength of the slabs, and the excessive loads applied to the building due to the modifications made by Sampoong. The building was also poorly constructed.

3.6.2 Legal Actions

An opinion column in The Korea Times stated "The owners did not bother to ask for approvals – especially as they knew the changes were against the building code. They simply bribed the relevant officials" (The Dawn 2004).

The final report was delivered by the Seoul District Prosecutors Office, entitled "The Final White Book of Finding Out the Real Truth of the Collapse of the Sampoong Department Store." The public was outraged. In particular, the news that the senior executives had fled the building without warning others was disturbing. The report on the collapse, as well as earlier structural and construction failures, suggested a widespread pattern of corruption in the country's construction business. A government survey of high-rise structures throughout Korea found 14 % were unsafe and needed to be rebuilt, 84 % required repairs, and only 2 % met standards. Joon Lee, the chairman of Sampoong, and his son Han-Sang Lee, were convicted and sent to prison for 10 1/2- and 7-year terms, respectively. Twelve local building officials were found guilty of taking bribes of as much as $17,000 (U.S. equivalent) for approving changes and providing a provisional use certificate (Wearne 2000, pp. 111–112).

The cause of the Sampoong collapse, then, was not a technical issue as much as outright fraud. The Korean construction industry, protected by government regulation from outside competition, had become complacent. Bribes were used to get around the usual government checks and balances that serve to protect public safety.

3.7 Rana Plaza Building (2013), Bangladesh

It is a tragic fact that a scholar of failures and forensics, unfortunately, can expect a constant supply of new material. Shortly before drafting this chapter, a building collapse in Savar, Bangladesh, near the capital city Dhaka, was reported. The collapse had some similarities to the Sampoong Super Store collapse.

The Rana Plaza factory building before the collapse is shown in Fig. 3.6. Five garment factories were working within the building at the time of collapse (17 Days 2013). The eight story building collapsed on April 24, 2013 (Bangladesh 2013). The collapsed structure is shown in Fig. 3.7.

Fig. 3.6 Rana Plaza factory building before collapse (Wikipedia Commons 2013)

Fig. 3.7 Rana Plaza factory building after collapse (Wikipedia Commons 2013)

The upper stories of Rana Plaza had been illegally constructed, without a permit. Questions quickly arose as to why the factories had even been operating. The building owner, Sohel Rana, had provided false assurances to the factory owners that the building was safe. Cracks were observed throughout the structure the day before the collapse. A reporter rushed to the building but was prevented by the owner's employees and local police officers from filming the damage. The employees were forced to report to work (Death Toll 2013).

"A giant crack formed on the side of the seventh floor of the structure, for instance, prompting warnings from engineers not to let workers inside" (Newcomb 2013). The factory owner attempted to flee, but was arrested near the border with India (Tears and Rage 2013).

The New York Times reported on 10 May 2013 that "The authorities in Bangladesh now say the building was illegally constructed, with permits obtained through political influence. The owner, Sohel Rana, now in jail, was illegally adding upper floors to the structure at the time the building collapsed, officials said . . .

The Rana Plaza disaster led to nationwide mourning in Bangladesh as well as outrage because it appears that the accident could have been averted. A day before the collapse, an engineer examined cracks in the structure and warned Mr. Rana, as well as owners of the garment factories, that the building was unsafe and should be closed. Instead, workers were told to come to their factories the next morning" (17 Days 2013)."

By 2 ½ weeks after the collapse, the death toll had risen to 1,053. In an eerie parallel to Sampoong, a single female survivor was found alive after 17 days (17 Days 2013).

It was later reported "The architect of the eight storey building that collapsed in Dhaka, Bangladesh, killing more than 500 people, has spoken out for the first time, telling The Daily Telegraph it was planned for shops and offices – but not factories… Massood Reza, the architect who drew up the plan for Rana Plaza in 2004, said he was 'asked to design a commercial shopping mall' with 'three or four storeys for a market and then the upper two storeys were for offices.'

He said: 'We did not design it for industrial use. At that time the garment belt was not there. There was no demand for industrial buildings. If I had known that it was to be an industrial building, as per the rules I would have taken other measures for the building.'

Other architects stressed the risks involved in placing factories inside a building designed only for shops and offices. The structure may not be strong enough to bear the weight and vibration of heavy machinery.

The government's official investigation on Friday suggested that generators placed on the roof to power the factories – along with the vibration of sewing machines used to make garments – all combined to trigger the building's collapse" (Bangladesh: Rana Plaza architect 2013)."

Unfortunately, it does not seem as if the authorities in Bangladesh have the resources to carry out a thorough forensic investigation. Moreover, in Bangladesh, the garment industry wields considerable economic and political power and in the past has proved to be adept at avoiding prosecution (Yardley 2013a).

Following the Rana Plaza disaster, the government of Bangladesh and industry leaders asked 30 professors from the Bangladesh University of Engineering and Technology to start inspecting buildings. The professors were expected to complete the inspections in addition to their normal teaching responsibilities. By July 2, 2013, they had inspected 100 buildings including 66 of Dhaka's estimated 1,500–2,000 garment factories. Two of the buildings were closed because of risk of collapse (Yardley 2013b).

3.8   Conclusions

All five of these cases were failures in service. In some cases, warnings of impending failure were not heeded. Each of the disasters let to considerable loss of life. Some of them also led to important changes in how the profession of engineering has been practiced in their respective countries.

3.8.1 Protecting the Public

What is the role of engineering ethics in protecting the public? In the United States, it has continued to evolve over the past decade. Engineering undergraduate accrediting standards under ABET, Inc. now require programs to demonstrate that students demonstrate "an understanding of professional and ethical responsibility" (Criteria 2012). Although ABET Inc. currently primarily accredits programs in the United States, it also has accredited programs in 25 other countries.

The ethical responsibilities of engineers are also governed by professional societies and licensing boards. Many state licensing boards now require 15 h per year of continuing education for licensed professional engineers, and some mandate that ethics be included.

Just as engineering ethics have evolved in the United States in the direction of greater protection of the public, it can be expected that such evolution will continue to occur around the world.

3.8.2 Disciplinary Mechanisms

As these cases illustrate, there are many different disciplinary measures that may be taken against engineers who fail in their duty to protect the public. These include criminal prosecution, revocation of licensure, and expulsion from professional societies.

In the United States, the civil engineering profession has traditionally been self-policing, and has relied on the latter two mechanisms. It could be argued that, by and large, these have been successful at protecting the public.

In other countries, enforcement of professional standards has relied more on criminal prosecution. Criminal prosecution is a blunt instrument, often unable to distinguish between cases where there was ample warning of impending disaster, such as the Vaiont Dam, and those where there was little or no warning, such as the Malpasset Dam. In cases of corruption, such as the Sampoong Super Store and the Rana Plaza, criminal prosecution appears to be appropriate.

3.8.3 Comparing the Cases

In all of the countries where these cases occurred, the profession of engineering is governed by laws and regulations. However, in countries where corruption is rampant, bribes may be used to circumvent the laws, weakening the protections put in place for public safety. In both the Sampoong Super Store and Rana Plaza collapses, the owners made deliberate decisions to increase their profits in ways that exposed the public to considerable risk.

In contrast, the other three cases occurred in nations where corruption was less of an issue. In those cases, the decisions made by engineers and owners seem more reasonable, at least on the surface. Questions remain as to what actions owners should take when there are indications of potential impending failure. In both the Malpasset and Vaiont Dam failures, there were indications of the disaster beforehand, but those warnings were not properly interpreted and heeded.

Three of these cases, the Hyatt Regency, the Sampoong Super Store, and Rana Plaza, led immediately and directly to a questioning of whether the engineering profession was doing enough to protect the public in their respective countries. In the Sampoong and Rana Plaza cases, further investigation revealed broad patterns of corruption in the building industries in South Korea and Bangladesh. Such introspection is valuable if it leads to improved practices.

The obligation of engineers to protect public safety is universal, and there is no dispensation for practicing in a less developed country. A human life has value whether in the United States, France, Italy, South Korea, or Bangladesh, and that fact needs to be a focus of engineering practice anywhere in the world.

References

17 days in darkness, a cry of "Save Me," and joy. (2013). *The New York Times*, May 10, 2013.

American Society of Civil Engineers (ASCE). (2013). *Code of ethics.* http://www.asce.org/Leadership-and-Management/Ethics/Code-of-Ethics/. Accessed 14 May 2013.

Bangladesh building collapse kills at least 70. (2013). *The New York Times*, Apr 24, 2013.

Bangladesh: Rana Plaza architect says building was never meant for factories. (2013). *The Telegraph*, May 18, 2013.

Criteria for Accrediting Engineering Programs. (2012). *Engineering accreditation commission.* ABET Inc., Oct 27, 2012. Baltimore, MD.

Death toll at 256 in Bangladeshi building ollapse. (2013). *The New York Times*, Apr 25, 2013.

Delatte, N. (2009). *Beyond failure.* Reston: ASCE Press.

Gardner, N. J., Huh, J., & Chung, L. (2002). Lessons from the Sampoong department store collapse. *Cement & Concrete Composites, 24,* 523–529.

Genevois, R., & Ghirotti, M. (2005). The 1963 vaiont landslide. *Giornale di Geologia Applicata I (Journal of Applied Geology I), 1,* 41–52.

Gillum, J. D. (2000). The engineer of record and design responsibility. *ASCE Journal of Performance of Constructed Facilities, 14*(2), 67–70. May 2000.

Hendron, A. J., & Patton, F. D. (1985). *The vaiont slide: A geotechnical analysis based on new geological observations of the failure surface.* Vol. I, main text, technical report GL-85-5, department of the Army, U.S. Army Corps of Engineers, U.S. Army Engineer Waterways Experiment Station, Vicksburg, June 1985.

Levy, M., & Salvadori, M. (1992). *Why buildings fall down: How structures fail.* New York: W. W. Norton.

Luth, G. P. (2000). Chronology and context of the Hyatt regency collapse. *ASCE Journal of Performance of Constructed Facilities, 14*(2), 51–61. May 2000.

Ministere de l'Agriculture (Ministry of Agriculture). (1960). *Final report of the investigating committee of the Malpasset Dam.* Translated from the French and published by the U.S. department of the interior, national science foundation, and Israel program for scientific translations, published in two volumes in 1963.

National Society of Professional Engineers (NSPE). (2013). *NSPE code of ethics for engineers.*http://www.nspe.org/Ethics/CodeofEthics/index.html. Accessed 15 May 2013.

Newcomb, T. (2013). Bangladesh catastrophe: Why concrete crumbles. *Popular Mechanics.* May 2, 2013.

Pfatteicher, S. K. A., & Pfatteicher, S. K. A. (2000). "The Hyatt Horror:" Failure and responsibility in American engineering. *ASCE Journal of Performance of Constructed Facilities, 14*(2), 62–66. May 2000.

Ross, S. (1984). *Construction disasters: Design failures, causes, and prevention.* New York: McGraw-Hill.

Rubin, R. A., & Banick, L. A. (1987). The Hyatt regency decision: One view. *ASCE Journal of the Performance of Constructed Facilities, 1*, 161–167.

Tears and rage as hope fades in Bangladesh. (2013). *The New York Times,* Apr 26, 2013.

The Dawn of modern Korea – collapse of Sampoong Department Store. (2004). Opinion. *The Korea Times.* Oct 14, 2004.

Wearne, P. (2000). *Collapse: When buildings fall down.* TV Books, L.L.C. (www.tvbooks.com) New York.

Wikipedia Commons. (2013). http://en.wikipedia.org/wiki/Main_Page. Accessed 14 May 2013.

Yardley, J. (2013a). Justice still elusive in factory disasters in Bangladesh. *The New York Times,* June 29, 2013.

Yardley, J. (2013b). After disaster, Bangladesh lags in policing its maze of factories. *The New York Times,* July 2, 2013.

Chapter 4
US Engineering Ethics and Its Connections to International Activity

Rachelle D. Hollander

Abstract This chapter situates the field of engineering ethics in the US as one among other related specialties in professional ethics, particularly biomedical and research ethics. International interest in engineering ethics and social responsibility has broadened the conceptual frameworks and methodological approaches in the field. The chapter describes efforts of US government agencies, professional engineering societies, and the US National Academies (NA) including NA efforts that involve similar bodies in other countries, to address ethics in science and engineering research and practice in the US and abroad and to stimulate consideration of ethics in engineering education. It also describes on-going international activities to develop an international code of research ethics as well as a wide variety of standards that affect engineering practice. These activities and efforts provide part of a global background for engineering ethics education in the future.

Keywords Engineering ethics • Engineering ethics education • Research ethics • US and international engineering ethics • Policy influences in engineering ethics education • Standard setting and engineering ethics in international contexts

4.1 Introduction

This chapter on international standards for engineering begins with a very brief overview of the development of engineering ethics and a list of some motivating questions for the field as it is approached in the US. The chapter highlights the differences between the fields of research ethics and engineering ethics, and notes how these differences play out to some extent in the international arena. A description follows of some of the ethically-relevant international activities in which US engineering is engaged, noting in particular their role in relationship to

R.D. Hollander (✉)
National Academy of Engineering, Washington, DC, USA
e-mail: RHollander@nas.edu

© Springer International Publishing Switzerland 2015
C. Murphy et al. (eds.), *Engineering Ethics for a Globalized World*, Philosophy of Engineering and Technology 22, DOI 10.1007/978-3-319-18260-5_4

ethics education in engineering. It postulates that the priorities for engineering ethics education should include greater and more systematic attention to these ethically relevant international activities.

4.2 Overview and Motivating Questions in US Engineering Ethics

The academic field of engineering ethics developed in the US in the early 1970s with other inquiry concerning issues of practical and professional ethics. Perhaps biomedical ethics was earliest to gain both scholarly and public interest; engineering and research ethics soon followed.[1]

Controversies concerning engineering catastrophes and research misconduct are likely to have fueled public demands and professional responses. There is a difference in motivation and response between the fields of engineering and research ethics. Governmental responses to engineering catastrophes required review and sometimes change of various regulations affecting safety.[2] Work in engineering ethics accelerated when ABET, the accrediting body for engineering programs at colleges and universities, initiated a requirement that engineering students demonstrate understanding of ethics for the profession and practice in 1985 and continues with the inclusion of ethics in ABET Engineering Criteria.[3] The response to research misconduct resulted in new Federal requirements concerning research integrity.[4] Current NSF and NIH requirements for ethics mentoring of postdoctoral students and ethics education for graduate and undergraduate students involved in research that has federal support have also stimulated engineering ethics research

---

[1]Observing Bioethics (OUP 2008) by J. Swazey and R. Fox provides a history and sociological analysis and contains interviews with founders of the field of bio- or biomedical ethics. In the mid-1970s a new program on ethics and science and technology at the National Science Foundation supported a conference drawing together teams of engineers and philosophers to stimulate research in the field of engineering ethics; this resulted in publications such as Martin and Schinzinger (1996). and Harris et al. (2014). The early history of research ethics as a field in the US may be said to begin with issues that arose with human subjects in medical experimentation; see Fried (1974), and to broaden to considerations about scientific misconduct in medical research that arose in the 1970s LaFollette (2000). There is significant overlap between the fields of biomedical ethics and research ethics.

[2]Numerous agencies in the federal government incorporate attention to issues of safety, e.g., the Department of Transportation has automotive safety as one of its fundamental purposes; the National Transportation Safety Board investigates many kinds of transportation accidents; the Consumer Product Safety Administration rules on safety issues on these products, ranging from lawnmowers to infant cribs.

[3]See J. Herkert on ABET's ethics requirement at http://www.onlineethics.org/cms/8944.aspx.

[4]See http://ori.hhs.gov/federal-policies.

and education.[5] While engineering expertise is involved in this process, the research integrity conversation primarily engages scientific associations, universities, and the funders and gatekeepers of academic research.

As the field of engineering ethics got underway, initial research involved philosophers, often with engineers as collaborators, and focused on ethical problems from the perspective of individual engineers. Ethical theory was used to provide useful conceptual clarification of the ethical dimensions of possible action. Codes of ethics and other guidance concerning human development and human rights from professional societies and national and international bodies, provided other resources, as did laws and other regulations. The field continues to draw on these resources.[6]

An important resource for the field is case studies, which take numerous forms and have numerous uses. The case descriptions, commentaries, and findings of the NSPE Board of Ethical Review are a rich source of material for engineers faced with and scholars wishing to examine ethical problems.[7] Cases can be hypothetical or historical, provide positive or negative role models, focus on everyday or rare and large-scale events, individual or organizational actions. They can take a prospective or retrospective view – that of the agent or the judge. They can describe value conflicts or problems of drawing lines between what is permissible, unacceptable, recommended, or forbidden. The cases can be simplified to illustrate a particular concept (called thin description) or illustrate real life messiness so as to demonstrate how people may legitimately arrive at different solutions. Finally, cases may illuminate a problem from the perspective of an individual engineer, or they may include documentation and analysis of issues that can only be resolved at an organizational or societal level.[8]

More recent research in the field of engineering ethics in the US includes international partners as well as historians and social and behavioral scientists as well as science and technology studies scholars and examines issues of complex

---

[5]NSF policy is found at http://www.nsf.gov/bfa/dias/policy/rcr.jsp. For NIH, see http://grants.nih.gov/grants/guide/notice-files/NOT-OD-10-019.html.

[6]See for instance the collection of ethics codes at http://ethics.iit.edu/research/codes-ethics-collection. Also, see the ongoing activities of the Scientific Responsibility, Human Rights, and Law program of the American Association for the Advancement of Science (AAAS) at http://srhrl.aaas.org/. For an example of another kind of resource see the Nuffield Council on Bioethics statement on biofuels at http://www.nuffieldbioethics.org/biofuels-0. Two publications from the National Academy of Engineering, National Academies Press focus on ethics education for graduate students in science and engineering. The titles are Ethics Education and Scientific and Engineering Research: What's Been Learned? What Should Be Done? (2009) and Practical Guidance on Science and Engineering Ethics Education for Instructors and Administrators (2013). Resources for engineering ethics education are also discussed in Chap. 4: NRC and NAE (2014).

[7]NSPE BER Cases: http://www.nspe.org/Ethics/EthicsResources/BER/index.html.

[8]Much of this delineation of types of engineering ethics cases is drawn from work by Ronald Kline, Bovay Professor, Cornell University. Many of these cases can be found in the Online Ethics Center at www.onlineethics.org. See also Kline (2010).

systems and collective as well as individual responsibility. Projects examine the role responsibilities of engineers in organizations, and the collective responsibilities of engineering societies and the organizations and networks that include and employ engineers and develop and promote and regulate engineering innovations. They also go beyond ethical questions about the impacts of engineering projects to study ethical aspects in project design and implementation, including public engagement. They ask whether traditional theories and approaches in ethics must be revised, augmented, or cast aside in light of the difficulties that complexity in sociotechnical systems and ecologies creates for fulfilling ethical responsibilities in project development and management.[9]

Some scholars believe that the increasing complexities require new ethical theories or concepts and approaches if they are to be resolved, while others hold that further elucidation of already extant understandings can handle the problems, but that new policies and practices may be required. This bifurcation in views seems common to many fields of science, engineering, and technology ethics.

Different cultural understandings about the nature and focus of engineering, and therefore of engineering ethics, add another dimension of difficulty to articulation as well as the discussion and resolution of specific issues. The literature of engineering cultures provides an introduction to this subject.[10]

All of these matters underlie the differences in emphasis and scope apparent in questions posed in engineering ethics education from those in research ethics. From the US perspective, general problems in engineering ethics would include:

How can the domain of professional engineering responsibility be legitimately circumscribed? Are there ethical commonalities covering all engineering fields, or is different guidance needed?

To what assistance in fulfilling their ethical responsibilities are engineers entitled? Again, does this differ between fields?

How can engineered systems identify and address issues of social and societal inequities? Who has responsibilities to do this; who shares these responsibilities?

How should engineers participate in societal determinations about promoting innovation or assessing and managing risk? Who should bear the costs and risk of failure? Are there ethically better and worse ways to distribute benefits or compensate those who are not benefited or are harmed? Who should decide?

Are there engineered systems that are too complex or dangerous to introduce in society, if the engineers' paramount responsibility for health, environment, and safety is to be maintained?

How should engineers and the engineering professions contribute to a future that is economically, environmentally, and socially sustainable?

---

[9]See Johnson and Wetmore (2007).

[10]Downey et al. (2005, 2007).

What is the proper role for engineering ethics across political, geographical, and generational boundaries?[11]

4.3 Relevant US Efforts

The subject matter in this book indicates just how complex all of these questions are. Sensitivity to the historical contingencies that make a difference to how engineering as well as engineering ethics get defined and evolve is important to this subject matter. The rest of the chapter describes some of the relevant historical efforts of US government agencies and professional engineering societies, and the US National Academies (NA), including NA efforts that involve similar bodies in other countries. This description indicates that US attention to research ethics emphasizes matters that are of lesser priority to engineers and engineering societies than are their own national and international activities that set standards that affect engineering practice. National and international standard setting activities are an important way in which engineers negotiate their commitments to public health, safety, and well-being, as well as sustainability; but these activities have not received much attention in engineering ethics education in the US.

The previous section noted a motivational difference between research ethics and engineering ethics, illustrated by the difference between ethics and catastrophes and ethics and research misconduct. The latter impelled US federal agencies to develop research integrity standards and ethics education requirements for research with Federal support. Engineers, like physicians and unlike many scientists, have historically had direct contact with publics and clients, and ethics is a priority in these situations. Today, if engineers and engineering feel distanced from approaches emphasizing research issues, that distance might be traced to the difference between ethics in practice and ethics in research settings. Much engineering research occurs in contexts of practice. The rollouts of new communications technologies and software provide examples.

US Federal Agencies The years from the early 1980s onward have seen a steady march of engagement of federal research and development agencies in defining ethics requirements for research. Perhaps among the earliest requirements, research with human and animal subjects provides a model that continues – with the responsibility for education and monitoring devolving to the institutions directly engaged in the research.[12] Research integrity requirements follow this model, with academic institutions holding the initial responsibility to assure that accusations about

---

[11] These questions can be found in National Research Council and National Academy of Engineering (2014). Chapter 4, 125–127 and in ch 2 – Goals and Objectives for Instruction, Kline (2013).

[12] The responsibilities of academic institutions for human subjects in research are found at http://www.hhs.gov/ohrp/humansubjects/guidance/45cfr46.html. A good discussion of ethics issues in animal research is at http://www.onlineethics.org/cms/13119.aspx.

fabrication, falsification, and plagiarism are attended to in ways that follow federal standards.[13] Insofar as engineers lead or participate in research, they must obey these requirements. The US National Institutes of Health (NIH) through the Office of Research Integrity (ORI) has issued requirements and guidance on research ethics and ethics education for institutions that receive NIH support, and the US NSF has done the same. ORI in particular has funded a number of conferences to examine and report findings from projects on issues of international research integrity.[14]

The previous director of the National Science Foundation Subra Suresh articulated well the federal government's interest in maintaining and promoting research integrity as requiring

> [an open flow of information with transparent processes to promote] rigorous peer review and scientific integrity.... The heads of major science and engineering research funding agencies from nearly 50 countries....took the first steps toward this goal by convening at the...NSF in Virginia, for the first Global Merit Review Summit. ...held on 14 and 15 May 2012, ...to ensure that science functions in a coherent and well-coordinated manner among developed and developing nations, maintains the public trust, and addresses each nation's unique needs for economic growth, national security, and human capital development.[15]

The attention continues to emphasize aspects that have priority for science, but perhaps less so for engineering, even though Suresh is an engineer. In addition, although it mentions the role of research integrity in maintaining public trust, the statement emphasizes research as enhancing wealth and security rather than as a societal investment that can stimulate goals of equity, community, public health and safety, or sustainability and environmental protection (although human capital development could contribute to these alternative goals).

The US National Academies The National Academies (NA) is a federally chartered private non-profit organization whose members provide advice to the nation about policy-relevant matters that involve science, engineering, and technology. Towards the end of the 20th century US federal agencies asked it to do a study of how to define scientific misconduct. This resulted in the 1992 publication *Responsible Science* which provided the first definition as "fabrication, falsification, or plagiarism."[16] A NA project began in 2011 to update this volume; 17 committee members provide broad and distinguished representation in the sciences and engineering as well as the field of science and engineering ethics; the chair is an engineer and one member is at a foreign institution. As the first report from the NA since the

---

[13]See footnote four for the electronic address for regulations on research integrity and the responsibilities of academic institutions. Ob cit?

[14]The third international conference on research integrity was held in Montreal Canada May 3–5, 2013. It developed a statement on research integrity. See http://www.wcri2013.org/overview_e.shtml.

[15]http://www.sciencemag.org/content/336/6084/959.full. Accessed 9-3-2013.

[16]NAS-NAE-IOM (1992).

influential 1992 volume, results from this effort are likely to be influential also. The focus will be on issues of research integrity rather than those of practice or social responsibility, however, so it is not clear that it will have much influence in engineering ethics education or over the practices of engineers or the organizations in which they operate.[17]

Another activity demonstrating this distance between research and engineering ethics is underway in the IAC and the IAP networks. As its press release states, "The IAP is a global network of the world's science academies, launched in 1993." The NA is a member of the IAP and of the InterAcademy Council (IAC), which was founded in 2000, to allow "top scientists and engineers around the world to provide evidence-based advice to international bodies."[18] IAC headquarters is in The Netherlands; IAP, in Italy. In 2012, with staff assistance from the NA, the IAP and IAC released a report to promote research integrity. This short policy report provides principles and guidelines for individual scientists, educators, and institutional managers, on research integrity, which includes addressing issues of research management, reward, principles, practices and culture throughout the world. The title is *Responsible Conduct in the Global Research Enterprise*. Its main recommendations focus on the conduct of research for individual researchers and research institutions as well as funding agencies and journals; however, this focus raises only a few questions about the social responsibilities of researchers or research institutions. The recommendations to funding agencies and journals are similarly focused.[19] Therefore these results may be of minor significance to engineers and engineering and engineering education.

The IAC is working to develop an educational guide, like the third edition of the booklet "On Being a Scientist" (OBAS) published by the National Academies Press in 2011.[20] The international guide will cover a wider range of topics (e.g. biosecurity) than OBAS and be aimed at a broader audience (not just graduate students and junior investigators but senior researchers, administrators, etc.), and also be explicitly international. Several staff members at the NA, in the Development, Security, and Cooperation (DSC) section of the Division of Policy and Global Affairs (PGA), are involved in this project, which has some funding from the Department of State. This report was originally scheduled for publication in 2013 and is nearing completion as this chapter goes to press.

Engineering Societies The prior efforts are attempts to establish standards for ethical behavior in research. With respect to practice many longstanding approaches speak to engineering concerns. Within the engineering societies, these efforts often

---

[17]Project information is available at: http://www8.nationalacademies.org/cp/projectview.aspx?key=49387.

[18]The IAP press release about the report is at www.interacademies.net/News/PressReleases/19784.aspx.

[19]InterAcademy Council/IAP (2012). Forward, p. v.

[20]National Research Council (2009).

take the form of developing and revising codes of ethics. As engineering codes of ethics have evolved over the years, they have begun explicitly to address social responsibility matters as well as matters of professional ethics. Ethical issues for engineering practice traditionally were concerned with relationships among professionals and are reflected in statements from early codes of ethics in the engineering societies, as versus more recent statements about issues that arise between engineers and society. Within the last several decades, for instance, more and more codes have incorporated the principle to protect public health, safety, and welfare as paramount. This principle creates a degree of protection for engineers who try to address health and safety issues that their superiors would prefer not to address. Even more recently, some codes have added requirements concerning environmental protection and sustainability.

Engineers in other countries share these ethical concerns. Besides the international sections and members in US engineering societies, there are equivalent international professional and other engineering organizations that have codes of ethics or other statements that express ethical positions.

The American Society of Civil Engineers (ASCE) has posted on its website information about the work of the engineering societies, through the American Association of Engineering Societies (AAES), an organization comprising more than a dozen engineering and related professional societies, to develop a code of ethics outlining fundamental principles for all engineers.[21] This effort continues, but there is no agreement as yet on an overall code. The website informs readers that, in comparing the codes of 11 member societies, the degree of similarity was found to be strong. Each society had requirements about competence and objectivity or honesty in its code. Two ethical issues closely related to competence and objectivity, namely, professional merit and conflicts of interest, also figure in all 11 of the AAES member codes reviewed. Another commonality concerns the public's health, safety, and welfare. Ten of the 11 AAES society codes state "the engineer's "paramount obligation' to public welfare, and all 11 require engineers to disclose adverse safety or health consequences arising in connection with their duties." Sustainability is becoming prominent in the codes of a number of these societies.

The information on the website notes differences between requirements in state licensing board codes of conduct and model rules drawn up by the National Council of Examiners for Engineering and Surveying (NCEES), and AAES codes of ethics. Only eight of the AAES member codes address bribery, fraud, and corruption; only six codes address the ethical ramifications of offering or accepting gratuities; only four concern themselves with contingency fees. Yet these are areas of significant international and widely shared concern.

Of significance in this discussion is the fact that, in the US, engineering fields distinguish between professional engineers and those sometimes called graduate

---

[21]http://www.asce.org/Ethics/A-Question-of-Ethics/2011/January-2011/.

engineers.[22] The former must be licensed and uphold professional standards or risk losing their licenses. Graduate engineers have earned degrees from accredited programs and do engineering work but have no state engineering license.[23] Furthermore, employees in companies with "engineer" in their job title need not have engineering training; this is a result of an industrial exemption in the engineering licensure, which allows the use of the term engineer but never professional engineer.[24] Members of all of these groups are included in the specialized professional engineering societies.

It is noteworthy that these codes do not single out research as a subject for ethical notice. They address issues of practice primarily. For the most part if not entirely any problems of research ethics could fall within these more general parameters. However, the lack of explicit attention indicates that the standard priorities in research ethics are not those of primary interest to engineering societies or their members.

There are many other ways in which engineers throughout the world address matters of ethics and professional responsibility, with and without labeling them ethics. New international groups such as Engineers Without Borders have sprung into being in order to put expertise at the service of groups that otherwise could not afford it.[25] US engineering societies work with each other, with societies in other countries, and with other governments and standard-setting organizations to develop, maintain, and modify technical standards that address issues of public health, safety, welfare, and environment. Examples range from long-standing concerns about such matters as corruption to quite technically and socially complex issues. Here are some examples.

A number of US engineering societies are working together and with international counterparts to address issues of bribery, extortion, and other corrupt activities that affect engineering practice. While this may be viewed as mundane, it is clearly important and in fact complicated internationally. The World Federation of Engineering Organizations (WFEO) is the international umbrella for associations representing individual professional engineers, and a number of US engineering societies are members. It represents associations from over 90

---

[22]There is a similar discussion of these issues in Chap. 4 NRC and NAE, *Emerging and Easily Accessible Technologies,* 2014.

[23]"What is a PE?" National Society of Professional Engineers (http://www.nspe.org/Licensure/WhatisaPE/index.html) Accessed on August 1, 2012; "Regulation and Licensure in Engineering," Wikipedia (http://en.wikipedia.org/wiki/Professional_Engineer) accessed on August 1, 2012; "Professional vs. Non-Professional Degrees," Washington University in St. Louis (http://ese.wustl.edu/undergraduateprograms/Pages/ProfessionalvsNon-ProfessionalDegrees.aspx) Accessed on August 1, 2012.

[24]"Signing Off on Engineering Documents," Online Ethics Center (http://www.onlineethics.org/cms/4606.aspx) Accessed on August 1, 2012.

[25]The website of EWB-International is http://www.ewb-international.org/. The website of EWB-US is http://www.ewb-usa.org/ Another example: Engineers for a Sustainable World: http://www.eswusa.org/.

countries, which themselves represent approximately eight million engineers. It has formed a Standing Committee which is tasked with promoting anti-corruption actions internationally. The Committee on Anti-Corruption (CAC) has the purpose of engaging the worldwide engineering community in the global efforts to fight corruption. It works with UN and other agencies and partners to address training needs for small and medium size businesses and to build capacity in civil society organizations. Information is at http://www.wfeo.net/stc_anticorruption/. As the site notes, all kinds of stakeholders have this issue on their agenda and the CAC seeks to learn about what they are doing and how they can join with them to address their needs, and how to encourage WFEO members to work with local engineers to help.

On a more cutting edge of ethically relevant activities fall issues of carbon management: With the belief that it is the responsibility of engineering professionals and engineers to respond to the global challenge of addressing worldwide greenhouse gas emissions, five major US engineering societies – the American Institute of Chemical Engineers (AIChE); the American Institute of Mining, Metallurgical, and Petroleum Engineers (AIME with AIST, SPE, TMS, SME); the American Society of Civil Engineers (ASCE); the American Society of Mechanical Engineers (ASME); and the Institute of Electrical and Electronics Engineers (IEEE), have joined together to organize an initiative to the challenges posed by developing and implementing technologies for Carbon Management. The website will serve as a resource on the challenges for their collective one million members and society at large; see http://fscarbonmanagement.org/content/education-and-outreach.

International Standard-Setting Engineering practice requires and relies on standards to meet its professional responsibilities. Ongoing international activities set standards that affect engineering practice worldwide; the International Organization for Standardization (ISO) oversees and undertakes these activities. ISO is an independent, non-governmental organization made up of members from the national standards bodies of 164 countries with a Central Secretariat in Geneva, Switzerland, that coordinates the system. See ISO information at http://www.iso.org/iso/home.html.

Founded in 1947, since then ISO has published more than 19,000 International Standards covering almost all aspects of technology and business. ISO standards cover risk, quality, environment, and energy and in 2010 it released a standard on social responsibility.[26] The standards provide useful direction to businesses and industry regarding good practice in all of these areas. Some of the standards include certification requirements, which are a strong incentive for good practice. The ISO 9000 family addresses various aspects of quality management and contains some of ISO's best known standards. They provide guidance and tools for organizations

---

[26]On candidates for standard setting for building inter-organizational trust to create efficient, effective, and ethical relationships, see Muskin (2000). Wikipedia has a useful basic discussion of the social responsibility standard (which doesn't include a certification requirement) at http://en.wikipedia.org/wiki/ISO_26000. It references Hahn and Weidtmann (2012).

who want to ensure that their products and services meet customers' requirements and will continue to improve. The ISO 14000 family addresses various aspects of environmental management. It provides practical tools for organizations looking to identify, control, and improve their environmental performance.

ISO 26000, the standard on social responsibility, provides guidance for businesses and organizations to use to develop good organizational governance and labor practices and address issues of fair operating practices and consumer protection and access to information, community involvement and development, and environment and sustainable development.

The American National Standards Institute (ANSI) is the US member of ISO. Comprised of government agencies, non-profit and for-profit organizations and companies, academic and international bodies, and individuals, the American National Standards Institute (ANSI) represents the interests of more than 125,000 companies and 3.5 million professionals. The ANSI website is at http://www.ansi.org/default.aspx. On September 13, 2013, the website posted a reminder about the upcoming 2013 annual US Celebration of World Standards Day. The importance ANSI places on the ISO process can be seen in how it characterizes what it does: "bring confidence to businesses, governments, and consumers, impacting reliability from farm to table, manufacturer to retailer, and workplace to home. Standards facilitate technological innovations and foster evolving social and environmental compliance practices that increase the health and safety of the world and its citizens. They are the foundation upon which developing nations build their economies and compete in the global market. Developed by collaborative bodies that act as forums for the public and private sectors to come together for a common good, standards evolve to meet the changing needs of a world at work and play."

ANSI and ISO charge for membership and also for some of their products and services, including information about standards. Helping organizations qualify for certification with respect to this family of standards is itself a lucrative business. Thousands of organizations in over 100 countries have adopted the quality standards, for instance, and according to ISO many more are in the process of doing so because the standards work, they save money, customers expect it, and competitors use it. Analysis of the ethical issues that these processes involve is overdue in engineering ethics and engineering ethics education.

4.4 Conclusion

Today's world is one of complex and inter-dependent socio-technical systems in which responsible human beings, organizations, and communities must develop and operate safe and reliable technologies ethically and transparently, so as to contribute to the health and welfare of society. Social justice and sustainability must factor into this equation.

This chapter presents a sketchy outline of engineering ethics as a distinctive academic field in the US, its somewhat rocky relationship with the field of research ethics, and examples of the kinds of engagement that US agencies and professional engineering societies and standards setting organizations have with international engineering ethics activities. The interactions include activities initiated in the US as well as by others. These relationships appear to be durable and to have good results. Whether they will have enough foresight to address increasing complexity and interactivity are open questions. Additionally in the US and abroad there is increasing attention to engineering ethics education. US activities have not yet benefited from sustained interactions with programs in other countries. No programs have paid much attention yet to the ethical implications of standard setting nor to the problems raised by corruption or other forms of interference with good practice in engineering. A systems approach to the development of engineering ethics education programs might address some important gaps concerning engineering ethics in this global context.

References

Downey, G. L., & Lucena, J. (2005). National identities in multinational worlds: Engineers and engineering cultures. *International Journal of Continuing Engineering Education and Life Long Learning, 15*(3–6), 252–260.

Downey, G. L., Lucena, J., & Mitcham, C. (2007). Engineering ethics and identity: Emerging initiatives in comparative perspective. *Science and Engineering Ethics, 13*(4), 463–487.

Fried, C. (1974). *Medical experimentation: Personal integrity and social policy.* Amsterdam: North-Holland. American distributor: American Elsevier Publishing.

Hahn, R., & Weidtmann, C. (2012). Transnational governance, deliberative democracy, and the legitimacy of ISO 26000: Analyzing the case of a global multistakeholder process. *Business Society* published online 21 Oct 2012. doi: 10.1177/0007650312462666

Harris, C. E. et al. (2014). *Engineering ethics: Concepts and cases* (5th ed.). Boston, MA: Wadsworth - Cengage Learning.

InterAcademy Council/IAP. (2012). *Responsible Conduct in the GlobalResearch Enterprise: A Policy Report.*

Johnson, D., & Wetmore, J. (2007). STS and ethics: Implications for engineering ethics. In E. Hackett (Ed.), *New handbook of science and technology studies* (pp. 567–582). Cambridge, MA: MIT Press.

Kline, R. (2010). Engineering case studies: Bridging micro and macro ethics. *IEEE Technology and Society Magazine, 29*(4), 16–19.

Kline, R. (2013). "Balancing priorities: Social responsibility in teaching responsible conduct of research" in National Academy of Engineering. *Practical guidance on science and engineering ethics education for instructors and administrators*: Papers and Summary from a Workshop Dec 12, 2012 (pp. 17–25). Washington, DC: The National Academies Press.

LaFollette, M. C. (2000). The evolution of the "Scientific Misconduct" issue: An historical overview. *Experimental Biology and Medicine (Maywood), 224*(4), 211–215.

Martin, M., & Schinzinger, R. (1996). *Ethics in engineering* (3rd ed.). New York: McGraw-Hill.

Muskin, J. B. (2000). Inter-organizational ethics: Standards of behavior. *Journal of Business Ethics, 24,* 283–297. Netherlands: Kluwer Academic Publishers.

NAS-NAE-IOM. (1992). *Responsible science: Ensuring the integrity of the research process.* Washington, DC: National Academy Press.

National Academy of Engineering. (2013). *Practical guidance on science and engineering ethics education for instructors and administrators.* Washington, DC: National Academies Press.

National Academy of Engineering. (2009). *Ethics education and scientific and engineering research: What's been learned? what should be done?* Washington, DC: National Academies Press.

National Research Council. (2009). *On being a scientist: A guide to responsible conduct in research third edition.* Washington, DC: National Academies Press.

National Research Council and National Academy of Engineering. (2014). *Emerging and readily available technologies and national security a framework for addressing ethical, legal, and societal issues.* Washington, DC: The National Academies Press.

Swazey, J., & Fox, R. (2008). *Observing bioethics.* Oxford; New York: Oxford University Press.

Chapter 5
"Global Engineering Ethics": Re-inventing the Wheel?

Michael Davis

Abstract This chapter defends four propositions: (1) There are already global standards of engineering practice, standards that are (more or less) what is needed; therefore, there is no need to create such standards from scratch. (2) Engineers do not need global registration or licensing. Neither registration nor licensing would increase the "professionalism" of engineers. Professionalism is independent of both registration and licensing. (3) Engineers have and, therefore, do not need, a global code of engineering ethics (though there is room for improvement). (4) Engineers have (and therefore do not need) a global curriculum for engineering ethics. Rather than trying to "re-invent the wheel", we should be proposing refinements to engineering's standards of practice, codes of ethics, and curriculum.

Keywords Profession • Discipline • Culture • Codes of ethics • State of the art

5.1 Introduction

My participation in this volume began with an email (May 23, 2012) inviting me to participate in an "International Workshop on Engineering Ethics for a Globalized World" the "specific themes" of which would "include the prospects for (1) international standards of engineering practice; (2) international standards of professionalism and registration; (3) an international code of ethics; and (4) an international engineering ethics curriculum." Judging both by that call, and the workshop itself, this volume seems to be inspired by at least four propositions: (1) that there are no international standards of engineering practice (or, at least, none suitable for a globalized world); (2) that engineers need international standards of professionalism, especially those imposed by registration or license; (3) that

M. Davis (✉)
Humanities Department, Illinois Institute of Technology, Chicago, IL 60616, USA

Center for the Study of Ethics in the Professions, Illinois Institute of Technology, Chicago, IL 60616, USA
e-mail: davis2643@gmail.com

© Springer International Publishing Switzerland 2015
C. Murphy et al. (eds.), *Engineering Ethics for a Globalized World*, Philosophy of Engineering and Technology 22, DOI 10.1007/978-3-319-18260-5_5

engineers need (and do not have) an international code of engineering ethics; and (4) that there is not now even a minimally satisfactory international curriculum for engineering ethics.

What I shall do here is challenge all four of these propositions. I shall do that not because I doubt this volume's usefulness, but because I believe re-thinking its inspiration would facilitate its work. The problem of cultural relativity, insofar as it exists, is not primarily a problem within engineering, that is, a problem to be solved by radically revising engineering's standards of practice or education to make them "truly global". The problem of globalism is, instead, primarily a tension between engineering and every culture in which engineers must operate—national, religious, corporate, and so on—a tension to be treasured rather than escaped. The chief revisions in engineering standards now needed are those clarifying engineering's critical stance toward other cultures, for example, a preamble stating that engineers should follow the same professional standards in Rome as in Beijing, the same in Buenos Aires as in Mumbai. That, anyway, is what I shall argue.

5.2 Culture

By culture, I mean any distinctive way that members of a group have of doing something. So, for example, classical music is one (musical) culture, the culture of classical musicians (and their listeners); jazz, another, the culture of jazz musicians. General Motors has one (business) culture; Apple, another. And so on. We may distinguish at least three (overlapping) kinds of cultural difference: manners, nomenclature, and technology.

By manners, I mean those ways of doing things that are indifferent until adopted as *the* way of doing that thing. For example, bowing is one way to greet; hugging, another; and shaking hands, a third. Every human association should have a way to greet but which of these ways (or ones much like them) it adopts does not matter. What are good manners in one setting may be bad manners in another. So, for example, in a society where bowing is the only form of greeting, a bear hug may seem threatening rather than friendly. I shall ignore differences in manners hereafter. The old saying seems to apply, "When in Rome, do as the Romans."[1]

By nomenclature, I mean what terms are to be used (the standards of naming a certain culture has established). Nomenclature is more than manners insofar as nomenclature itself matters (whether morally, economically, or some other way important to members of the culture). So, for example, giving measurements

---

[1] This use of "manners" is stipulated; it does not claim to cover all senses of "manners" (though I do think it catches the main sense of "mere manners"). If "manners" (in another sense) involves, say, harmful treatment of another, then it is not manners in my sense. We would need a new term (perhaps "custom" or "style"). I do not claim my description of culture is complete, only complete enough for my purposes here.

in metric rather than English is, though merely a matter of nomenclature, not indifferent (hence, not a mere difference in manners). The metric system has advantages the English lacks. Not only is metric more likely to be understood everywhere (an extrinsic advantage), it is also easier to use (an intrinsic advantage).

By technology, I mean not artifacts as such but artifacts embedded in the social network that designs, builds, distributes, maintains, uses, and disposes of them, and the standards governing that network. Technology matters insofar as it creates wealth, improves living conditions, or otherwise adds to the good things in life—or destroys wealth, worsens living conditions, or otherwise adds to bad the things in life. Nomenclature is only about language; technology is about things (as well as language).

Technological culture can vary quite a bit, even within a given locale.[2] So, for example, the Chicago Transit Authority (CTA) has one system for collecting fares (certain plastic cards, card readers, and so on); Chicago's Metropolitan Transit Authority (METRA), another; the national passenger trains running through Chicago (AMTRAK), yet another. A CTA rider who takes a METRA train for the first time will have almost as much difficulty buying a ticket as he might had he instead tried to buy a subway ticket in London, Stockholm, or Tokyo.

5.3 Culture and Globalism

Given this way of understanding culture, engineering is undoubtedly itself a culture, that is, engineers have a distinctive nomenclature and technology, what I usually call "the discipline of engineering" (distinctive names for tools, processes, tests, and the like, and distinctive ways of keeping records, designing artifacts, testing materials, and so on). This culture is (in large part at least) what distinguishes engineers from other technologists, such as architects, computer scientists, and synthetic chemists (Davis 1998).[3]

Engineering's culture is international. That internationality is what allows an engineer in Japan or North American to read without difficulty (once translated into English) the design, documentation, or schedule prepared by an engineer in Europe or China. It is what allows an Egyptian or Brazilian engineer who speaks English to teach in an Irish school of engineering or work in an Australian engineering firm.

Given the scale of international trade today, it is not surprising that engineering is an international discipline. What is surprising is how long engineering has been such a discipline. For example, when the U.S. Military Academy at West Point began

---

[2] Is technology a feature of every culture? Not every culture. We can imagine humans so primitive that they lack any artifacts at all yet have distinctive ways of singing, dancing, or doing other things.

[3] See esp. Chaps. 1, 2 and 3. While all cultures must be disciplines, I am not sure that all disciplines must be cultures. Perhaps there can be idiosyncratic disciplines (for example, a certain way I train myself to walk even though no one else walks that way).

teaching engineering soon after its founding in 1802, it used French textbooks—and soon had a French military engineer on its faculty. The same would have been true in Spain or Russia at that time (Davis 1998, pp. 18–19).

There is, of course, a distinction to be made between "international" and "global" culture. To be international, a culture need only cross one international border, say, the U.S.-Canadian border. To be global, the culture must be the same everywhere on earth (or, at least, almost everywhere), from the most developed country to the least. What I now want to argue is that engineering is global in this sense—and useful largely because it is.

Consider: Why might a country without engineers want them—want them enough to hire them from outside or establish an engineering school of its own with much the same curriculum as engineering schools elsewhere? The answer cannot be that, without engineers, there can be no technology. No country is so undeveloped as not to have its own craftsmen, inventors, and the like who can repair old artifacts or build new ones. Indeed, such local technologists are more likely than engineers to create artifacts appropriate to the locale. What engineers offer, what they have offered for at least a century, is technology made according to international standards—so that, for example, a part cast in Thailand will fit into an engine that, though assembled in Brazil, is composed of parts made in Germany, Mexico, and Canada.[4]

A country that does not want to participate in international trade in this way is free to develop its own curriculum to train "quasi-engineers", "technical managers", or whatever it chooses to call its own brand of technologist. But, insofar as its schools train its local technologists to standards different from those to which engineers are trained, engineers (strictly so called) will have difficulty using the work of those local technologists. The country will be shunning the international culture in which engineers work.

5.4 A Ulysses Contract

When a would-be employer makes being an engineer a condition of employment (or just employs an applicant in part because she truthfully declared herself to be an engineer), that engineer is obliged to bring engineering's international culture into that workplace. Indeed, to employ an engineer is to enter (something like) a "Ulysses contract" concerning engineering culture.

---

[4]This is, of course, true only generally or for the most part. Chicago's failure to have one system of collecting fares for its three main forms of rail service (mentioned earlier) is proof enough that engineers do not always achieve universal standardization. Though there are historical reasons for this (for example, a failure to foresee the need to integrate the three systems until recently), engineers—though perhaps not everyone else—would seek to eliminate it.

Recall Ulysses' ingenious solution to a problem he faced while trying to return home after the Trojan War. He wanted to hear the Sirens' song. The only way to hear it was to sail close to their rocks. Yet anyone hearing the Sirens became temporarily mad. Since Siren-induced madness generally caused sailors to wreck themselves upon the rocks, Ulysses did not want his sailors to hear the Sirens or to obey him while he listened. He solved the problem by having his sailors block their ears with wax, tie him to the mast, and leave him to listen, gesture, rant, and struggle until the ship was again in safe water. Though he remained captain of the ship, Ulysses gave up the right to give new orders until he could no longer hear the Sirens' song. A Ulysses contract is an arrangement by which someone freely gives up certain rights in order to protect himself from his own poor judgment (Homer 2014).

To employ an engineer is (in part) to employ someone who should *dis*obey certain orders an ordinary employee should obey. For example, to employ an engineer is to employ someone who is (in part) supposed to insist on standard safety factors even when the employer orders him to use a less demanding standard (and even when doing as ordered is both legal and likely to benefit the employer). For the ordinary employee, loyalty entails giving the employer the benefit of the doubt on such questions; for an engineer, it does not. Employers nonetheless employ engineers for certain work instead of other plausible candidates, such as industrial designers or technical managers. They do that for at least one of three reasons:

First, experience has taught employers that they should not depend on themselves or their ordinary employees (of whatever rank) in such matters. They have found an engineer's ways of doing such things to be more reliable than their own or that of other employees.

Second, employers have reason to believe that suppliers, customers, consumers, or the public will trust them more if they are known to do as engineers recommend than if they are not so known. They employ engineers to have the benefit of the good reputation that engineers have earned by adhering to engineering culture. Part of what engineers sell is trustworthiness of a certain kind.

A third reason employers may employ engineers is that a law requires engineers rather than others to make certain decisions. The law requires that because experience has taught lawmakers that engineers are better trusted with such decisions than others are. An engineer who puts engineering standards before her employer's wish, rule, or apparent welfare is not being disloyal to her employer, paternalistic, or otherwise overstepping her bounds—no more than Ulysses' crew were when they ignored his Siren-induced gestures. The engineer is protecting the employer from itself, giving it what it bargained for, the benefit of engineering's discipline (Walker 2012).

5.5 Global Standards

It follows from what I have just said that engineering has a global culture, its own ways of doing things wherever around the globe those things are to be done. It does not follow, however, that engineering has exactly the right ways of doing such things. Since new global standards are adopted every year—by ASME, IEEE, ISO, and so on—, that conclusion is not troubling. The important question is not whether engineering needs new standards (of course, it does) but what new standards it needs. My answer is that any needed changes in standards need not be large (though some may be). More important for our purposes, there is no need to create global standards of professionalism to replace local standards because global standards already exist, no need to create a global code of engineering ethics because that too already exists, and no need to create a global curriculum for engineering ethics because even that already exists. I shall now defend this answer.

By professionalism, I mean practice according to professional standards. A "true professional" (in this sense) is a member of a profession who acts as members of that profession are supposed to act. Professionalism has nothing to do with license or registration.[5] It is not surprising, then, that Canadian engineers, who must be licensed, do not seem to show more professionalism than Dutch engineers, who live under a government that has no system for licensing or registering engineers. The claim that professionalism requires licensing or registration relies on a mistaken understanding of professions, one I have criticized at length elsewhere (Davis 2009).

By ethics, I mean any morally permissible standard of conduct that all members of a group (at their rational best) want all others to follow even if the others' following the standard would mean having to do the same. Given this definition, engineering ethics includes not only engineering's code of ethics (so called) but also its technical standards—for example, those concerned with details of safety, quality, documentation, and sustainability—*provided* the standards are morally permissible and what all engineers at their rational best want all other engineers to follow even if that would mean having to do the same.

Is that proviso ever satisfied? That is an empirical question about which it would be good to have more research. But I do think that, absent clear evidence to the contrary, there is good reason to believe not only that (most) engineering standards are morally permissible but also that they are what engineers would (at their rational best) endorse. If there is any standard that is clearly morally wrong, let it be pointed out. I am sure most engineers would willingly see it repealed.[6] Meanwhile, engineers have good reason to want their present standards followed (until changed).

---

[5]Licensing and registration (where they exist) seem to be forms of consumer protection or government oversight (just as dog licenses and vehicle registration are). In general, licensing settles who can legally do certain things (those with the appropriate license) and who cannot (those lacking the appropriate license). Registration, in contrast, generally creates a list of those who can claim a certain title (for example, "registered engineer"). Registration does not settle who may legally do certain things and who may not.

[6]Those standards that some engineers think immoral while others do not are a bit harder to deal with. They have the advantage of being in place—and a presumption in their favor because of that. We might require some time and a good deal of discussion before we could decide whether, supposing there are any such standards, engineers at their rational best would accept them (or reject them) or must instead recognize disagreement concerning them to be one of those disagreements reason alone cannot settle.

Those standards are what distinguish engineers from other technologists (architects, chemists, computer scientists, and so on). They are what make it possible for engineers to make a living as engineers (rather than as technologists of another sort).

Do engineers nonetheless lack a global code of engineering ethics? The answer may seem to be, "Obviously, yes." After all, there are many codes of engineering ethics: ASME has one code; the European Federation of National Engineering Associations (FEANI), another; and so on.[7] These codes seem to differ in geographical origin, style, language, and substance. None seems explicitly designed for use in developing countries.

But that "Obviously, yes" rests on at least two mistakes. First, it assumes that no code of engineering ethics is global when *most* are, that is, they apply to "engineers" as such, not merely to members of an association (as the IEEE code does), to a geographical division of engineers (as the Asian Declaration of Engineering Ethics does), or to engineers working in certain places (say, less developed countries).[8] So, most codes of engineering ethics at least claim to apply globally. Second, that "Obviously, yes" assumes that existing codes, whatever their global pretentions, differ so much they cannot jointly guide. They cancel each other out. Yet, the codes do not differ much, if at all, in substance. Even differences that at first seem large generally disappear upon inquiry. For example, engineers following a code of ethics without a provision on sustainable development seem to interpret the environmental or public-welfare provision in their code to include it.

But what, it might be asked, are we to make of differences that do not disappear upon inquiry? Judging by history, I would say we should suspend judgment. Some differences disappear over time, for example, as one branch of engineering after another comes to see the importance of the new standard in question. There is, of course, the possibility of a fundamental difference between branches of engineering, one that remains generation after generation (though I know of none now). But, were there such a difference, especially on anything engineers regard as important, engineering might split—as, for example, "scientific management" split off from engineering almost a century ago.

---

[7] The first association listed here was *formerly* known as the American Society of Mechanical Engineers. For a large selection of engineering codes, see: http://ethics.iit.edu/ecodes/ethics-area/ 10 (accessed June 13, 2012). To be charitable, I am ignoring the Model Code adopted by the World Federation of Engineering Organizations in 1990, though it was meant to guide the writing of codes of engineering ethics in "all nations". http://www.wfeo.net/about/code-of-ethics/ (accessed October 10, 2012). Thanks to Jun Fudano for reminding me of this document to me.

[8] Might there, despite the wording, still be an *implicit* limitation of ASCE's code to civil engineers, of ASME's code to mechanical engineers, and so on? Perhaps. But there are at least three reasons to doubt it. First, the codes are so similar that there is little point to such a limitation in practice. Second, engineers do not seem to regard distinctions between fields of engineering in the same way they regard the distinction between, say, chemistry and chemical engineering. They do not talk as if belonging to another field of engineering would excuse conduct their code declares unethical. Third, there does not seem to be any positive evidence of that implicit limitation. Absent evidence to the contrary, we should take the codes at their word.

I therefore suggest that we think of the many formal codes much as we think of the many dictionaries of American English. Though they differ, they (more or less accurately) report the same underlying reality. One code may omit what another includes because of a different purpose, format, publication date, or the like, not a difference concerning the underlying reality. The variety in formal codes is consistent with (more or less total) agreement on the "unwritten code".[9]

So, there is already a global code of engineering ethics. But is it adequate for use in developing countries? After looking through proposed alternatives, I think the answer is plainly yes. Consider, for example, the best alternative I've found, Harris's "Guidelines". One Guideline suggests: "Respect the cultural norms and laws of host countries, insofar as this is compatible with the other Guidelines." (Harris 2004, p. 516).[10] As I understand the Ulysses contract into which the host-country (or, rather, the specific employer) enters by employing an engineer, the engineer is under an obligation *not* to respect the culture of the host-country insofar as that geographical culture is inconsistent with engineering's culture (something about which Harris' Guidelines are silent). Indeed, the engineer is there (in part) to do what a fully acculturated local would not do. This way of thinking about the relation between host-country culture and engineering culture may seem less surprising if we consider one of Harris's own examples (nepotism) (Harris 2004, p. 517).

An engineer is told to hire several assistants to help oversee a building project. Ordinarily, she would hire the engineers who seem most likely to do a good job. But her employer suggests one of his relatives, noting (correctly) that looking after family is part of "host-country culture". An engineer faced with the same question in Chicago (her "home country") would know what to do. She would say,

> You can hire whom you want. But, if you want me to take responsibility for the hiring, you must let me choose those I think will do the best job. One thing I must consider is that, if I have to fire an assistant for incompetence or laziness, it will be harder if he is one of your relatives. Hiring your relative would bias my judgment in a way that might affect the public health, safety, or welfare. I won't do that.

Now, contrary to what Harris's Guidelines suggest, I think an engineer should say precisely the same in a distant host-country, however family-centered and less developed. She was given the job of selecting her own assistants because she is best placed to choose well—or because the employer wanted the benefit of her reputation

---

[9]The unwritten code may be amended either by an informal shift of usage or by a formal decision of suitably important engineering associations.

[10]The only competitor to Harris' proposal in the existing literature is in Luegenbiehl (2004, 2010). Of course, there is always the possibility of a set of standards I have overlooked or that someone will someday write. I cannot disprove the possibility that that one will overcome the objections made here. All I can claim is that I am entitled to my conclusion until that other standard is produced.

for being so placed. She cannot choose well if she is required to use criteria likely to interfere with choosing well. If the employer wants someone to do what any other employee would, he should not have assigned that responsibility to an engineer.[11]

5.6 Curriculum

I therefore think that a good "global ethics curriculum" for engineering would look much like the ethics curriculum we now have. There could be a few improvements, of course, but these would be useful at home as well as abroad. I shall mention three.

The first improvement is explicitly rejecting the claim that engineering ethics as understood in the U.S. must end at its borders—or, at least, at the borders of the "developed world".[12] This relativist claim should be rejected because it relies on the unstated premise that geographical culture takes precedence over technological culture. I have yet to see an argument for that claim. In addition to the arguments already made against it here, I offer this one:

The Asian Declaration of Engineering Ethics (2004) was adopted by the chief engineering societies of China, Japan, and Korea. It is unusual in being explicitly limited to a geographical area (Asia) and in being the work in part of "developing countries" (China and Korea). It seems to have been conceived as a way to document differences between Asian and non-Asian standards. Yet, the only significant difference between the Declaration's standards and American or European standards *seems* to be the last: "[Asian engineers shall]:... Promote mutual understanding and solidarity among Asian engineers and contribute to the amicable relationships among Asian countries."[13] Since all engineers should, I suppose, promote mutual understanding and solidarity among engineers everywhere and contribute to amicable relations among all countries, the Asian Declaration's last provision is no more than a special case of what is (or should be) a more general obligation, one that could be included in any code of engineering ethics. Even that last provision does not reveal a significant difference in geographical culture within engineering.

The second improvement in curriculum I would like to see is increased use of examples drawn from outside the U.S. I would like that in part because I am tired of the small store of American examples that most engineering ethics texts

---

[11]Harris' solution was to accept one relative but only one, assuming (I suppose) that saying yes once would make it easier to say no next time. Yet, my experience is that agreeing to dubious conduct once makes it harder, not easier, to refuse the next time. The first time means there is a precedent for doing it next time.

[12]See, for example, Harris (2004), p. 504; "there are important social and cultural differences in host countries that can affect the way an essentially U.S.-based code can be applied in host-country environments."

[13]http://ethics.iit.edu/ecodes/node/5076 (accessed May 28, 2012).

share. Last year I used a text by several Dutch philosophers—written in English—and enjoyed discussing the European examples (van de Poel and Royakkers 2011). My students had no trouble appreciating them. The European examples helped my students understand engineering as a global profession (while also teaching them how to practice in the U.S.). The addition of examples from less-developed countries would do the same.

The third improvement in curriculum I would like to see is more explicit discussion of the two senses in which engineers seem to use "state of the art". In one, a piece of technology is state of the art if it is "the most advanced" (that is, has all the "bells and whistles" money can buy). In another sense, a piece of technology is state of the art if it gives the best fit between budget, conditions, purpose, and engineering standards. State of the art in the first sense is an esthetic criterion, not properly an engineering criterion at all. Insofar as engineering is supposed to improve the material condition of society, only technology that is state of the art in the second sense is good engineering. So, for example, a complex system of pumps for keeping a high-rise's basement dry might be state of the art in the first sense, that is, exceed U.S., European, or Japanese standards in impressive ways, but not be state of the art in the second sense because the system will not keep the basement dry in the less-developed country in which it will be installed, a country where electrical power is off several hours a day almost every day. State of the art in the second sense means giving a country without reliable electric power a technology appropriate to it—whether that means a backup of batteries or a large diesel generator or, instead, a high-rise without a basement, or no high-rise at all. The idea of state of the art as the best engineering solution is, of course, an idea that applies everywhere, not just in less developed countries.

My hope is that this volume, re-conceived in this way, will generate ideas as important for engineering in the most developed countries as in less-developed ones, smaller ideas perhaps, but also more useful.

References

Asian Declaration of Engineering Ethics (2004). http://ethics.iit.edu/ecodes/node/5076. Accessed 30 April 2015.

Davis, M. (1998). *Thinking like an engineer: Essays in the Ethics of a profession*. New York: Oxford University Press.

Davis, M. (2009). Is engineering a profession everywhere? *Philosophia, 37*, 211–225.

Harris, C. E. (2004). Internationalizing professional codes in engineering. *Science and Engineering Ethics, 10*(10), 503–521.

Homer, *Odyssey*, Bk XII: 165–200, http://www.poetryintranslation.com/PITBR/Greek/Odyssey12. htm#_Toc90268047. Accessed 9 July 2014.

Luegenbiehl, H. (2004). Ethical autonomy and engineering in a cross-cultural context. *Techné, 8*, 57–78.

Luegenbiehl, H. (2010). Ethical principles for engineering in a global environment. In I. van de Poel & G. David (Eds.), *Philosophy and engineering*. Dordrecht: Springer.

Van de Poel, I., & Royakkers, L. (2011). *Ethics, technology, and engineering: An introduction*. Chichester: Wiley-Blackwell.

Walker, T. (2012). The ulysses contract in medicine. *Law and Philosophy, 31*, 77–98.

Chapter 6
Engineering Decisions in a Global Context and Social Choice

Noreen M. Surgrue and Timothy G. McCarthy

Abstract In this paper we specifically tackle the methodological problem of constructing an ethical framework for how professionals are to work within the global setting, and while this is an issue for a number of disciplines and professions including medicine, computer science, and education, for purposes of this paper we turn our attention to engineering. Specifically, we examine engineering in a global context and offer an analysis of how conflicts between normative constraints, imposed from several points of view, on engineering solutions are to be resolved.

Keywords Reflective equilibrium • Philosophy and Linguistics • Normative constraints • Engineering ethics • Sending country • Receiving country • Rawls

6.1 Introduction

There are normative constraints on developing an ethical framework for any profession working in a global context (Hanson and Rothlin 2010; Yuksel and Murat 1999). These constraints are imposed from several points of view (Kapstein 2001; Steers et al. 2010). Resolution of the conflicts between competing normative constraints is a methodological issue requiring adjudication and resolution. The first we call the *sending normative constraints*: these are the constraints arising from the

Many of the ideas for this paper arose from research funded by The Hewlett Foundation, The Sloan Foundation, The National Institutes of Health, and The Robert Wood Johnson Foundation. We would like to thank Paolo Gardoni, Colleen Murphy, and Dana Beth Weinberg for their comments.

N.M. Surgrue (✉)
Women and Gender in Global Perspectives, University of Illinois at Urbana-Champaign,
910 South Fifth St., Champaign, IL 61820, US
e-mail: nsugrue@illinois.edu

T.G. McCarthy
Departments of Philosophy and Linguistics, University of Illinois at Urbana – Champaign,
810 S. Wright Street, 105 Gregory Hall, Urbana, IL 61801, USA
e-mail: tgmccart@illinois.edu

basic scheme of intrinsic values of the society from which the professionals come. In most instances the sending normative constraints will be those of developed nations. The constraints of the second sort we call the *host normative constraints*. These are the requirements arising from the values and belief systems within the host country, the country that is the target of the professional intervention. For the most part, the host countries are middle income and developing nations. Finally there are the *professional normative constr*aints, these are constraints imposed by a specific profession's norms, responsibilities, and limitations.

In this paper we specifically tackle the methodological problem of constructing an ethical framework for how professionals are to work within the global setting, and while this is an issue for a number of disciplines and professions including medicine, computer science, and education, for purposes of this paper we turn our attention to engineering. Specifically, we examine engineering in a global context and offer an analysis of how conflicts between normative constraints, imposed from several points of view, on engineering solutions are to be resolved.

The plan of this paper is as follows: first we offer some general remarks on the foundations of the enterprise and we flesh the characteristics of the three types of constraints; and we argue that none of these constraint types systematically holds sway over the others by suggesting the possibility of cases in which each sort of constraint is able to dominate the other two. We next turn to a detailed presentation of three case studies designed to underscore the contrastive interplay among the three types of constraints. We then outline a methodological framework for resolving conflicts among these constraint types based upon a generalization of Rawls-Harsanyi reflective equilibrium. Finally, we assess the case studies in the light of this framework.

6.2 Normative Constraints on Design Decisions

The engineering decision context is subject to three contrasting sets of normative constraints.[1] We shall call constraints of the first sort design constraints. These are the desiderata generated by the engineering context itself. They include constraints of economy, effectiveness, robustness, replicability etc. Secondly, there are the sending-normative constraints mentioned above: these are conditions generated by the social preference ordering of the intervening agents. These are not the design constraints but the conditions imposed by the preference ordering among social outcomes induced by the basic scheme of intrinsic values of the intervening society. It is natural to suppose that these are supervenient on the preference schemes of individuals across the intervening society, but such use as we shall make of this assumption is eliminable.[2] Finally, there are the host-normative constraints, which are the direct analogues, for the host society, of the guest normative constraints.

---

[1] By a normative constraint on an engineering outcome we mean an overall condition of acceptability for that outcome.

[2] For a classical statement of the assumption see Harsanyi (1955).

The difficulty we are addressing in this paper arises from the fact that there can be significant divergences between these sets of constraints and between the outcomes they generate; we will illustrate this in detail below. The problem will be to devise a rational scheme that achieves equilibrium between them. There is, for example, an exact symmetry in the ways in which the host- and guest-normative constraints are characterized, but since there is normally wide divergence between the social preference orderings of the host and the guest societies, the outcomes they generate can significantly diverge. We will see various examples of this sort of divergence below. The intervening society is typically a liberal democracy constrained by norms of social welfare, personal liberty, and capitalism, undergirded by a system of constitutionally protected rights. The social preference ordering flowing from this situation may differ strikingly from that of the host society, which may reflect an entirely different mode of social organization. Thus, for example, the intervening agents may confront a quasi-feudal host society constrained by religious norms. It is clear that these normative constraints can conflict. How are we to proceed in such a situation?

We can initially discard three hopeful thoughts, which are that one among these three sets of normative constraints will uniformly control the engineering outcome. The alternatives are these:

Pure paternalism: The normative constraints of the senders alone, in conjunction with the design constraints, determine the acceptable engineering outcomes.

Pure tolerance: The normative constraints of the hosts alone, in conjunction with the design constraints, determine the acceptable engineering outcomes.

Pure efficacy: The design constraints alone determine the acceptable engineering outcomes.

Paternalism and tolerance fail for symmetric reasons; both social preference orderings, that of the host agents and that of the intervening agents, count for something. Suppose, for example, the guests have no particular interest in preserving a species of inedible mushroom, the main habitat for which lies in the most direct path of a proposed highway. But these mushrooms are sacred objects for the hosts. The design solution is constrained to build the road around the mushrooms. It is clear that this same sort of example also defeats pure efficacy. At the same time, the guest-normative constraints count for something in the overall normative context of the design project. The most direct route for the proposed highway may take it straight through a pariah village; the host norms are indifferent to the destruction of the village, but the guest norms dictate that the highway is built around it. In general, the normative framework of an engineering intervention is a superposition of three (sometimes quite contrasting) sets of constraints.

The examples just rehearsed indicate that in situations of asymmetrical indifference the committed norms, either of the hosts or the guests, may dictate an adjustment in an engineering outcome that is optimal from a design point of view. This is sufficient to show that neither the host nor the guest norms exclusively control the engineering decision context. We shall now extend these considerations by arguing that these constraint-complexes can pairwise conflict, and that each complex properly controls in some situations but not in others.

First, sending-normative constraints can override both design-constraints and host-normative constraints. Let us suppose, in the example of the highway above, that there is a cost-sharing arrangement between the hosts and guests that would require a significantly larger contribution from the hosts if the highway were to be routed around the pariah village. The social preference ordering of the hosts may positively prefer the outcome resulting in the destruction of the village. The senders would be justified, from their own moral point of view, in implementing the more costly plan; and this would also seem justifiable from an external point of view. This concern would normally override the social preference of the hosts for the more efficient solution arising from their indifference to the destruction of the village.

Second, host-normative constraints can override design constraints and guest-normative constraints conjointly. Suppose that it is morally problematic from the senders' standpoint that a certain host subpopulation is socio-economically privileged, but that respecting those privileges is a necessary condition of getting anything done at all. The intervening guests may acquiesce to consequences they find morally repugnant as the price of being able to execute the project. The loss of the entire project may be less acceptable overall than infringing the relevant guest-norms as, for example, when loss of the project would results in a catastrophic outcome for the host society.

Finally, the design constraints supplied by the engineering context can also override both host- and sending- normative constraints. Suppose, for example, that a uniquely determined design solution forestalls a catastrophic environmental outcome, but that this solution has consequences which are morally regrettable from both the host and the guest points of view. Altering the previous case, suppose that implementing the design solution requires an outcome, such as the destruction of an innocent village, which infringes the well being of innocent individuals; but this time suppose that the hosts as well as the guests find this morally problematic. Let us suppose, however, that failure to implement the solution in question would result in the destruction of the society. For that reason the engineering solution seems acceptable from an external point of view, but systematically infringes both host and sending norms. The general rule suggested by this sort of case is that the design constraints override when failure to implement one of the engineering solutions compatible with those constraints would result in an existential threat to the host society.

6.3   Three Case Studies

In the first case the host normative constraints act as the controlling framework in which an engineering solution is chosen. Such a case is likely to involve a clash of cultural, religious, social, political, or gender based norms as between the host and guest. An engineering solution that meets the normative constraints of the host country may not necessarily be the best engineering solution but it is a compromise that allows an engineering intervention to be implemented against the background of the social preference ordering of the host society.

Example 1 Within country X there is a need for clean water, and in one particular village, C, the need is particularly acute. Within the village, the infant mortality rate is far above 300 per 1,000 live births, the leading cause of death among children under the age of 5 is diarrheal disease, there is no local source of clean water, and what water is brought into the village is not separated by usage: thus drinking, cooking, cleaning, and sewer water are one in the same. There are many non-governmental organizations (NGOs) with staff working in the village and one of the NGOs, whose mission it is to improve the lives of people by increasing access to clean water, has hired an engineering firm from country B in order to determine where a clean water well should be built and to build it as quickly as possible. The commitment to build the well and to bring in an outside engineering firm with expertise in well building has the support of the village leaders.

The choice of where to drill the well is dependent on a number of engineering factors including the ability to transport the equipment necessary to build the well and the water quality. Upon arriving in the village, the first task facing the engineers was to determine where to build the well. They spent 4 weeks doing technical assessments to determine where the well should be located. Engineering criteria, alone, were used to determine the site selection. Upon presenting the best possible site to the NGO staff and the village leaders, the village leaders protested and said the well could not be built. It turned out that the best site, as determined solely by engineering standards, involved bringing in equipment over the center of sacred grounds as well as digging at the outer perimeter of those grounds. The village leaders would not be moved by the design constraints. To them the respect required for that sacred ground was more important that the lives of the villagers; this was non-negotiable.

At that point, the engineers and the NGO staff conferred and asked if they could present an alternative sight in 2 weeks; the village leaders agreed. The engineers pointed out to the staff of the NGO that if they went with a different site the cost would be about 20 % higher, and maybe more. In addition, it would take an extra 60–90 days because of the routes available for bringing in the required equipment. The staff at the NGO understood that they were going with a more expensive and less efficient plan for building the well but decided that if a suitable second choice site could be found it was worth the additional expenditure of resources; a second choice site was the only way these villagers would be given access to clean water. The engineering team complied, a second choice site was determined, and the well built.

The second example situates the normative constraints of the sending country as the controlling framework in which a given engineering solution is implemented, against the background of all other constraints. In the case where the sending country's normative constraints override all other constraints, there is again likely to be a conflict among social, political, or religious norms with the host country; but in this case the impasse is broken in favor of the norms of the sending country.

Example 2 Company K, has been contracted to work on a project in country D. The project is to design and build a clinic for women and children. The clinic that is being built is to be a prototype and if it can be successfully built, within budget, officials in country D plan to commission the building of up to 75 more clinics whose purpose is to care for and improve the health and welfare of women and children. Company K employs the world's leading expert in clinic design and she is uniquely qualified to manage this project. Moreover, this particular female engineer has been denied other opportunities to work abroad because too few men in developing nations were willing to "take orders from a woman". The company realizes that if it is to maintain its reputation as a good place to work and continue to hire the best engineers possible, it must support its female worker force and not replace a female employee with a male employee, especially one less qualified, when a client, for reasons of bias, requests it. Company K's senior management team has promised that assignments will be made based on performance, without regard to gender; this promise was made to the senior staff as well as to the professional staff. It is a promise that the female staff takes very seriously: any breach of this promise would compromise company K's ability to recruit and retain the most talented professionals.

Company K has put together its best engineering and design team and is ready to begin the technical assessment and develop a timeline for the project in country D. When the design team arrives national and local leaders from country D meet them. When asked where the team leader is, the woman in charge asserts that she is the team leader and is excited to be working with them on this important project. Taken aback, a spokesperson for country D's leaders point out that it is impossible for a woman to be in charge and running a project. Because the contract calls for hiring as many local workers as possible, this would mean that local men would report to a woman and be supervised by a woman; this, the company officials are told, is completely unacceptable. Officials in company K state that the woman they have chosen to run the project is in charge and there will be no change. When the leaders of country D insist they will not work with a woman in charge, company K cancels the contract and departs country D.

This case illustrates the possibility of an irresolvable impasse between the values of the sending and host countries. Unless one side or the other is willing to compromise its basic scheme of intrinsic values, the engineering intervention will not be implemented. In this case company K faces living up to its commitment to and promise of gender equality and equity, and thereby maintain its ability to recruit and retain the best engineers possible or lose one contract because the country with whom they entered into the contractual arrangement does not believe a woman should be a leader or supervise local male workers. The firm, company K, decides the only thing that it can, and ought to, do is withdraw from the contract with country D. Therefore, in this case the implementation of an engineering solution is in effect cancelled by the normative constraints of the sending and host contexts.

In the third example we find the controlling normative framework to be the engineering design specifications. The norms governing an engineering intervention or solution, ceteris paribus, include conservation of resources, profit maximization, technological feasibility, efficiency, economy, robustness, and adaptability, and sustainability. These normative constraints set the context and boundaries within which the possible engineering solutions are debated and decided.

Example 3 A bridge engineering firm, Y, has been hired by country A. Y is being contracted to build a bridge that will serve as the main thoroughfare between the east and west portions of country A. It is anticipated there will be significant traffic, including very large loads on oversized trucks because this bridge will be the only route for transporting building materials needed for the planned economic development of the country. The bridge is to be built near a small village.

The firm, Y, has come in and completed its technical assessments and is ready to present its plan to the relevant decision makers within country A. Because of the location of the bridge national officials have asked key local leaders to be participate in the discussion of the firm Y's proposal. Without providing too many details, the proposed bridge will feature a number of triangles and have a height taller than anything else in the country. When the design is presented, no person from country A who must sign off on the proposal is willing to do so as it appears that the triangles and the height are violations of deeply held superstitions. The engineers decide that they must confront the superstitions head on and explain why there is no other design that will hold the loads projected for the bridge. They explain that there is no engineering firm in the world that will build the bridge with any different design given the intended use of the bridge. It is explained that if the design is not used the bridge will fail and many people will die. The country's leaders are also told that if they do not accept the design specifications company Y will not build the bridge, and they are sure no other engineering firm would either. After a long deliberation the decision-makers in country A agree that the design presented must be used and commit to educating the people in their country; they are not willing to forego having the bridge built.

6.4 Superposition of Norms

We propose to achieve a coherent equilibrium between the normative constraints associated with the sending society, the host society, and the design-context by utilizing a generalization of the Rawls' original position construction (Rawls 1999).[3] At a first approximation, knowing the guest norms, the host norms, and

---

[3]The familiar idea has various historical antecedents in Hume, Kant and Adam Smith, but its (now canonical) formulation appears in Rawls (1999). In effect, what is attempted below is a generalization of the notion of reflective equilibrium, which as originally understood was a sort of back-and-forth method for adjudicating conflicts between normative judgments of a given society, to a notion of equilibrium that adjudicates conflicts between norms of divergent societies and exterior constraints supplied by the engineering context.

the design norms, and equipped with the supposition that he is a member of one of the societies but not knowing which society or which individual role he plays, the question is what design solution an ideally rational agent would choose. We shall now attempt to make this rough suggestion more precise. We then conclude by arguing that the construction presented here affords a reasonable basis for resolving the conflicts at issue in the above examples.

Let us imagine Karl to be an ideal engineer, which from the point of view of any particular example means that he has complete knowledge of the design principles relevant to the problem at hand. We further suppose Karl to be logically omniscient with respect to the extraction of consequences from this information.[4] Karl is also idealized in two other respects: first, he has complete knowledge of both the sending and the host societies, including their associated social preference orderings and the normative constraints they generate, with the proviso that the particular identities of the individuals in these societies are masked. In particular, then, Karl cannot identify himself in the composite social situation comprising both the host and the sending societies. Finally, we suppose that Karl is a rationally self-interested agent. In this position, Karl is presented with the following decision problem: given that he is a member of the composite situation (but he doesn't know which individual role he plays), what, if any design solution would he choose for the problem at issue?

A number of aspects of this proposal should be clarified before we proceed to apply it. One aspect that needs emphasis here are the differences between the present construction and the version of the veil of ignorance deployed by Rawls, differences grounded in a corresponding disparity between the problems the constructions are designed to solve. In Rawls's theory one has a single social scheme and knowledge is provided to a cognitively ideal, rationally self-interested agent of the condition of each individual in society but as above the identities of the individuals are masked. The problem is to infer an impartial specification of norms of justice for the society, essentially a scheme of intrinsic values underlying a just social preference ordering. The agent knows he will occupy some position, but because the identities of individuals are masked he doesn't know which.

In the present construction, there are two social groups, one intervening in the other for a particular purpose that we assume at the outset to be sanctioned by both social preference orderings. The problem is to characterize the particular intervention(s) that are optimal with respect to exterior norms of justice. For this we suppose a cognitively ideal, rationally self interested agent, whom we were calling 'Karl', equipped with all relevant technical information as well as a complete specification of the condition of each individual in both societies, again with the identities of the individuals masked. Against the backdrop of the assumption, mentioned above, that the social preference orderings for the two societies are

---

[4]We can implement this idea in various ways. One model of this sort of cognitive ideality is to picture Karl as equipped with an 'oracle' that informs him, for any given yes-no question of the design context, whether it has an answer of the basis of the input principles and if so what that answer is.

completely determined by the preference orderings of the individuals within them, the assumption that Karl has complete information about the attitudes of each individual across the two societies entails that Karl also has complete information about the social preferences of the two societies.[5] The problem is then for Karl to determine, given that he is a member of the composite society (but he doesn't know which), what the optimal interventions are from his own point of view. The masking constraint is intended to ensure that the resulting optimal solutions are impartial, and in this respect is analogous to its function within Rawls's scheme; it is intended to ensure that the choice he makes from his individual point of view is also optimal from the composite social point of view.

Secondly, a slight clarification is perhaps required with respect to the question of what Karl is choosing between. Above we spoke of Karl as choosing between designs, or engineering interventions. But in normal social choice theory the direct objects of choice are overall composite social outcomes for the two societies. Such a social choice problem presupposes a demarcation of a range of possible outcomes, generated by the conditions of the problem at hand. However, given such a demarcation, to say that Karl "chooses" a particular intervention is to say that the intervention is realized in *each* situation that is an optimal choice for Karl within the range of possible social outcomes provided by the demarcation. Thus although there may be more than one optimal total situation, there may yet be only one optimal design solution.

However, there is one technical obstruction to the present proposal that does not confront Rawls. In the present construction we are, in effect, superposing two societies. In either version of the veil of ignorance we want to equip the agent in the choice situation with the assumption that he has an equal chance of playing the role of any individual. In Rawls's scheme this effect can be achieved by a simple statistical interpretation, where in effect the role of the determining agent will be determined by drawing a name out of a hat, one entry for each individual. But in the suggested variant we can't give a simple statistical interpretation of this sort, since the populations of the two societies normally differ in terms of cardinality. To equip Karl with the information that he has an equal chance of assuming the role of any individual in either society, we need only suitably weight the probabilities that he assigns to assuming the role of an individual in the host society, on the one hand, or in the sending society on the other. Thus, for example, if there are N members of the host society, M members of the guest society, and $N < M$, we arrange a lottery wherein each member of the host society is represented by M tickets in the hat, and each member of the sending society by N tickets, $2MN$ tickets in all. It is this statistical model that Karl believes will generate his role in the superposition of the two societies.

---

[5] If in fact this supervenience assumption about social preference is false, we will need to add to Karl's information a complete specification of the relevant social attitudes in both the host and the sending societies.

6.5 Assessing the Cases

Let us apply the suggested decision framework to the case studies above. We shall pursue the assessment at a qualitative level, but translating the narrative evaluations into either game-theoretic or decision-theoretic formal evaluations is a routine chore.

In the first case, the choices are between building the well in the initially recommended location, building it in the alternative location, and not building the well at all. If the first solution is implemented, and Karl plays the role of a member of the host population, some of his most important desires will be severely violated (sacred grounds will be desecrated), but others will be satisfied (there will be clean water). If he is assigned to the sending society his utilities will probably be undisturbed or marginally enhanced (if he is a member of the sending society that cares deeply about the problem). If the second solution is implemented and Karl is assigned to the host population, many of his most important desires will be satisfied and none will be violated; if he is assigned to the sending population, his preferences will be undisturbed or at worst marginally violated (if he is concerned about the additional cost associated with the alternative location). If the well is not built at all, Karl's desires are massively violated if he is assigned to the host society, and are at least somewhat violated in case he is a member of the sending society who cares about the problem the intervention is designed to solve, and are otherwise flat. Given this decomposition of the outcomes, it is clear that Karl will choose the second option if he regards himself as having an equal chance of playing any role in the composite situation.

In the second case, it is somewhat unclear what the range of possible choices should be taken to be. If the outcome in which the hosts compromise their gender prejudices is ruled out in advance, as being barred, in effect, by a state of nature, there are two possible outcomes, the one in which the senders compromise their egalitarian values and the clinic is built under the direction of male leaders and one in which the clinic is not built at all. If the outcome in which the senders don't compromise is ruled out, you get a symmetric pair of possible outcomes. If both compromises are ruled out, there is nothing to choose between; there is only one possible outcome. The most natural construal is the maximal one, where the choices are between building the clinic with a massive compromise of the values of the host society, building the clinic with a massive compromise of the values of the guest society, or not building the clinic at all.

The second thing one should notice about this example is that it is ambiguous what the proper outcome should be. In the story the intervening agents refuse to compromise their egalitarian program and the clinics are not built. Is this really the appropriate outcome? If we imagine that the project could be accomplished, though perhaps not equally well, by other means, or if the hosts could easily be persuaded to compromise, then the decision is justified if the egalitarian norms are of sufficient importance in the sending society. If, on the other hand, the proposed intervention is the *only* means of ensuring the building program, and there is no hope of persuading the hosts to compromise, and the forfeiture of that program has very serious health consequences for the trajectory of the host society, then it is far from clear that the decision to cancel the intervention is sound.

The present ambiguity is reflected exactly in the suggested model. In the first case, where other solutions are available using other companies and male supervisors, then within the class of possible choices we will find not only the situations that result from the proposed specific intervention but also the situations resulting from other possible effective interventions. To optimize his welfare in the composite situation, Karl will choose the one, or one of the ones, in which the building program is satisfactorily executed with little or no harm to the sending society; optimal among these are situations in which an alternative intervention team pursues the project, and the original prospective interveners decline. In such a situation, if Karl is projected into the host society, he stands a good chance of benefitting significantly from the intervention; and if he finds himself a member of the sending society he is essentially unscathed by the pursuit of that intervention. A similar story holds in case the hosts can be persuaded to compromise: in this case, among the possible alternatives are situations in which a program is mounted to do that. In a similar way Karl will choose the one, or one of the ones, in which the host society is persuaded to compromise and the intervention is successfully carried out. In both of these optimal scenarios the original interveners are justified in refusing to compromise their egalitarian norms.

In the second example, however, there are just two situations to choose from, the situation in which the putative interveners refuse to compromise and the clinics are not built, and the situation in which they compromise and the clinics are built, to the great benefit of the host society. In this case, as a prospective member of the host society (under the veil of ignorance), Karl is much better off in the second situation. As a prospective member of the sending society, Karl stands a small chance of being negatively affected in the second situation (if, for example, he is directly involved with the original intervention team), but his susceptibilities seem marginal in comparison with the first situation. If, as is assumed by the model, Karl regards himself as equally likely to find himself in either situation, he will choose the second. In the second example, then, in each admissible outcome the original interveners cave. And so there is a conflict between the conclusions delivered by the model in this example, because of an ambiguity in the specification of the choice set. That ambiguity reflects a corresponding ambiguity in the background conditions of the example itself.

When we examine the third example, we find that it largely coincides with a specification of the second. The possibility was raised of injecting into the choice set for example 2 situations in which a sustained attempt is made to talk the hosts out of their prejudices. The present example follows out a scenario of this kind, in comparison to the situation in which the interveners simply withdraw. The space of alternatives is then comprised of the situation in which nothing is done and the bridge is not built, and two situations in which a serious attempt is made to alter the relevant beliefs in the host community, one in which the attempt works, and one in which it fails. As a prospective member of the host society (under the veil of ignorance), it is clear that Karl will judge himself to be much better off in the second situation, in which the persuasive attempt is successful. As a prospective member of the sending community, his level of satisfaction is essentially flat as between the three scenarios if he is not concerned with the intervention, and markedly higher in the second situation if he is. On the construction of suggested model, then (again, since under the veil of ignorance Karl believes he has an equal chance of finding himself in either society), Karl will opt for the second situation. And that was the situation that seemed the right outcome in this case.

6.6 Conclusion

Our model correctly resolves the three case studies presented, not only in the sense that the case studies with determinate outcomes are assigned the proper results by the model but also in the sense that a case with an ambiguous outcome is not assigned a determinate outcome and the basis of the indeterminacy is explained by the model. Similar results may be obtained by applying the construction to other cases described more cursorily in the paper. What these findings suggest is that the construction of an ethical framework for engineering in a global context requires a set of metrics that allow for determining outcomes that unambiguously present themselves but also guide decision-making in the more likely scenarios involving multiple interpretations and highly ambiguous choices.

References

Hanson, K. O., & Rothlin, S. (2010). Taking your code to China. *Journal of International Business Ethics, 3*(1), 69–80.

Harsanyi, J. (1955). Cardinal welfare, individulistic ethics, and interpersonal comparisons of utility. *Journal of Political Economy, 63*(4), 309–321.

Kapstein, E. B. (2001). The corporate ethics crusade, foreign affairs, http://www.foreignaffairs.com/articles/57242/ethan-b-kapstein/the-corporate-ethics-crusade.

Rawls, J. (1999). A theory of justice. Cambridge, MA: Harvard university press, revised (Ed).

Steers, R., Sanchez-Runde, C., & Nardon, L. (2010). *Management across cultures*. Cambridge: Cambridge University Press.

Yuksel, O., & Murat, G. (1999). The globalization and global ethics: The case of less developed countries. http://www.opf.slu.cz/vvr/akce/turecko/pdf/YukselhuO.pdf

Chapter 7
Engineering Responsibility for Human Well-Being

Charles E. Harris, Jr.

Abstract Engineering codes and other authoritative engineering sources implicitly claim that engineering promotes human "well-being," but the term is not defined. Philosophical attempts to define well-being can be divided into desire-fulfillment (or preference) theories, hedonistic (or mental state) theories, and objective list theories. Psychological accounts of well-being are either hedonic or eudemonic. Both the objective list theory and the eudemonic theory consist of a set of goods that comprise well-being, and the lists have some similarity. The Capability Approach (CA) can also be interpreted as positing a set of conditions necessary to achieve well-being. I propose a reorganization of Martha Nussbaum's well-known list of capabilities and show their relation to engineering. The design criterion that CA suggests is that technology should enhance the capabilities of those who use it to achieve a life that they have reason to value. Some of the most noteworthy attempts to design for well-being have been directed toward the special needs of people in developing societies. The focus on designing for human well-being, which I call aspirational ethics, suggests that, in addition to following rules that prevent harming the public, engineers should develop certain virtues. In designing for developing societies, the most important virtues are empathy and compassion. For developed societies, the most important virtues are concern for the environment, sensitivity to the effects of technology, especially on society and human relationships, and creativity.

Keywords Well-being • Welfare • Quality of life • Aspirational ethics and codes

7.1 Introduction

Engineers generally recognize a responsibility to avoid unprofessional conduct, such as accepting bribes, having undisclosed conflicts of interest, and practicing outside their area of expertise. Let us call this category of professional obligations "Prohibitive Ethics." Engineers also usually recognize an obligation, within

C.E. Harris, Jr. (✉)
Texas A&M University, College Station, TX, USA
e-mail: e-harris@tamu.edu

C. Murphy et al. (eds.), *Engineering Ethics for a Globalized World*, Philosophy of Engineering and Technology 22, DOI 10.1007/978-3-319-18260-5_7

reasonable limits, to take due care to prevent harm to the public from engineering work. This obligation requires engineers to anticipate problems and to alert the proper authorities to actual or potential harms, and in extreme cases resort to "whistleblowing." Let us call this "Preventive Ethics." Engineering codes also appear to impose a third category of obligations on engineers: an obligation to promote the public good, presumably through their professional work. Let us call this category of more positive obligations "Aspirational Ethics." Compared to the other categories, this aspect of professional obligations is less clearly stated in the codes, and its meaning is much more controversial. In this essay, I explore this third category of professional obligation.

7.2 Interpreting the Codes

Virtually all engineering codes make reference to the "welfare," "well-being" or "quality of life" of the public. Here are some examples. The Preamble to the code of the National Society of Professional Engineers (NSPE) gives as the basis of such professional virtues as honesty and fairness the assertion that "engineering has a direct and vital impact on the quality of life for all people" (NSPE 2007). Accordingly, engineers "must be dedicated to the protection of the public health, safety, and welfare." Section I.1 says engineers are required to "hold paramount the safety, health, and welfare of the public." In III.2.a, engineers are encouraged to "work for the advancement of the safety, health, and well-being of their community" (NSPE 2007).

The code of the Association for Computing Machinery (ACM) obligates its members to "contribute to society and human well-being" (I.1) The same section says that "human well-being" includes a safe natural environment, on both the local and global levels (ACM 1992). The code of the American Society of Civil Engineers (ASCE) affirms that engineers should utilize "their knowledge and skill for the enhancement of human welfare and the environment.", and it holds that engineers should "recognize that the lives, safety, health and welfare of the general public" depend on the professional activities of engineers. Later in the same canon (Guidelines to Practice under the Fundamental Canons of Ethics, Canon 1), engineers are encouraged (not required) to "work for the advancement of the safety, health and well-being of their communities and the protection of the environment through the practice of sustainable development" (ASCE 2008, 13–14). Finally, the introductory statement of the code of the Institute of Electrical and Electronics Engineers (IEEE) says that its members recognize "the importance of our technologies in affecting the quality of life throughout the world" (IEEE 2006, 1).

The ambiguities in these statements with regard to well-being fall along two dimensions. First, it is not clear whether the terms "welfare," "well-being," and "quality of life" are synonymous, or are intended to convey different ideas. I shall assume that they are more or less synonymous, and I shall often adopt the term "well-being" to cover all three expressions. Second and more importantly, the terms are never defined, and even the general direction of their definition might be considered in doubt. Some statements point in a negative and preventive direction, such as the NSPE code's reference to an obligation to the "protection" of public welfare. Other statements point in a more positive and aspirational direction, using such terms as "work for," "contribute," and "enhancement." Do engineers have an obligation to merely protect the well-being of the public or do they have a more positive and robust obligation to promote human well-being?

A case for the more negative and preventive interpretation can be made from Case 82–5 of the NSPE's Board of Ethical Review (BER 1982). Here the Board addresses the situation of an engineer who was terminated for repeatedly protesting his employer's actions with regard to a defense contract, because he thought they resulted in excessive costs and delays. Citing the reference to "welfare" in section III.2.b of the NSPE code, the Board ruled that the engineer had a right "as a matter of personal conscience" (p. 3) to protest his employer's actions, but not a professional obligation to do so. In this case, which is the only BER case I have been able to find that cites the term "welfare" or synonymous terms, protecting the "welfare" of the public is protecting the economic interests of the public by preventing a wasteful use of public funds. Here we find not only an essentially negative and preventive interpretation of engineering obligation, but a narrow economic interpretation of "welfare."

A more positive interpretation of engineering obligation regarding human welfare is not only suggested by some statements in contemporary codes, but by earlier codes as well. Going back to a much earlier statement, the 1828 charter establishing the Institution of Civil Engineers in the United Kingdom (UK) defines engineering as 'the art of directing the great sources of power in nature for the use and convenience of man." At the time of the writing of this code, the expression "use and convenience of man"(Mitcham and Munoz 2010, 3–5) was often associated with utilitarian thinking and thus implied an obligation to maximize the good, so the expression suggests that engineering should promote the human good in a larger and more positive sense.

Recent statements from the National Academy of Engineering (NAE) of the United States also require a positive interpretation. At the end of the twentieth century, the NAE initiated a project to select the twenty greatest engineering achievements of the twentieth century. In an unpublished speech at the annual meeting of the NAE on October 22, 2000, President William A. Wulf, described the criterion for selection as "not technological 'gee-whiz,' but how much an achievement improved people's quality of life." He went on to say that the achievements selected are "a testament to the power and promise of engineering to improve the quality of human life worldwide" (Wulf 2000). Achievements cited by the NAE were electrification, the automobile, the airplane, water supply and distribution, electronics, radio and television, agricultural mechanization, computers, telephony, air conditioning and refrigeration, highways, spacecraft, the Internet, imaging, household appliances, health technologies, petroleum and petrochemical technologies, lasers and fiber optics, nuclear technologies, and high performance materials. In the *Foreword* to a book describing these achievements published by the NAE, Neil Armstrong describes the criterion for selection as those engineering achievements that had the 'greatest positive effect on mankind.' (Constable and Somerville 2003, vii) I conclude that, while prohibitive and preventive ethics is overwhelmingly the dominant theme in engineering codes, some statements inside and outside contemporary codes require a more positive and aspirational interpretation. It is not unreasonable to say, therefore, that the engineering profession has imposed upon itself an obligation to promote human well-being.

7.3 Accounts of Well-Being

But what does well-being mean? The most obvious answer is that it simply means the possession of the material conveniences and comforts of the kind cited in the NAE's list of the twenty greatest engineering achievements of the twentieth century. If one enjoys clean water, automobiles and the other conveniences on the NAE list, she by definition has a high state of well-being. This interpretation, however, faces difficulties. Since technological innovations are subject to rapid change and the NAE list will quickly be outdated, "well-being" cannot be identified with the NAE list or any other particular list in a straightforward way.

More importantly, some skepticism about the ability of technological progress to enhance well-being is justified. While little empirical evidence exists about the relationship of technology to well-being, there is empirical work on the relationship of wealth to happiness. What has come to be known as the "Easterlin Paradox" is the apparent contradiction in the empirical evidence that seems to show that, on the individual level, rich people are happier than poorer ones, but that, on the country level, people in countries that are getting richer are not getting happier (Easterlin 1974). More recent research has continued to confirm that as income increases, happiness tends to remain unchanged. In a more recent paper, Easterlin explains this phenomenon by arguing that as income rises, material aspirations do as well, so that people are not happier (Easterlin 1995). By contrast, non-pecuniary aspirations in such areas as health and family relationships do not continue in their ever-upward spiral. Since happiness is inversely proportional to the gap between expectation and reality, happiness is not increased with rising income. Luis Angeles explains the paradox by observing that a mere 3.6 % of the differences in happiness scores can be explained by income, the rest being due to such factors as employment, health, and marital status (Angeles 2011). If these non-pecuniary factors decline (e.g. by an increasing divorce rate), they can easily offset the positive effect of increasing income. Both authors agree that non-pecuniary factors are important in happiness.

As indicated, using this evidence to support the claim that technology does not increase well-being proceeds on some assumptions. First, since the studies by Easterlin and Angeles cited above focus on economic advance rather than technological advance, one must assume that the two are correlated. This has not been established. Second, in these studies, the kind of well-being studied is "happiness" in the sense of subjective well-being, a topic to be explored shortly.

I shall argue that this is an inadequate account of well-being. So it is probably safe to say that one should be wary of any simple or straightforward identification of technological advancement and increase in well-being.

Making further progress in understanding well-being requires a deeper look into the nature of the human good, especially as it is addressed in philosophy and psychology. Space permits only the briefest summaries of the central tendencies of research in these two disciplines.

The philosophical discussion of well-being has been dominated by three theories: desire-fulfillment (or preference) theories, hedonistic (or mental state) theories, and objective list theories (Adler 2012). For preference theorists, well-being is having one's preferences fulfilled. Mental state theorists hold that well-being is having desirable mental states. Objective list theorists maintain that well-being is the possession of some set of goods that are seen as intrinsic constituents of well-being. Objective list theorists differ in their list of goods, but they agree that the goods cannot be reduced to the mental states or preference satisfactions of individuals.

The critical discussion of these theories comprises a vast literature, but I believe the direction of the discussion can be summarized in the following way. Mental state theories have several flaws, one identified in Robert Nozick's well-known "experience machine" argument. Let us suppose that our brains are connected to electrodes so that we can stimulate them to produce any experience we want. According to Nozick, most rational people would reject this mode of life (Nozick 1974). Many philosophers conclude that the thought experiment shows that something matters (and hence is important for well-being) besides mental states.

One of the strengths of desire-fulfillment theories is that they can explain motivation to achieve well-being, since people are generally motivated to fulfill their desires. However actual preferences cannot be simply equated with well-being either, because (as philosophers tend to agree) an acceptable account of well-being must have critical force. It must, that is, be possible for a person to be mistaken about whether she has attained well-being, and this would not be possible if well-being were merely identical with actual preference satisfaction. This requires equating well-being only with more rationally considered preferences. Preferences must not be naïve and untutored but rather the result of fully informed rational deliberation—what Richard Brandt has called cognitive psychotherapy (Brandt 1998). Matthew Adler pushes such an idealized preference account still further toward an objective list account by requiring that the notion of well-being be the product, not simply of the deliberation of an individual, but of some measure of agreement between individuals who have engaged in rational and informed deliberation. While individuals may disagree to some extent, there may well be, Adler argues, a "zone of consensus" between the products of individual rational and informed deliberation that represents a satisfactory account of well-being (Adler 2012). The problem remains, however, that desire-satisfaction theory is presented on an abstract level and provides no indication of the content of rationally-considered desires. Furthermore, Adler avers that Martha Nussbaum's well-known list of capabilities —which we shall consider in more detail later—may be the best account so far devised of what rational people would desire (Nussbaum 2000, 78–80; Adler 2012).

This brings us to objective list theories of well-being. As would be expected, various lists have been proposed. Setting aside for the moment Martha Nussbaum's list, Philip Brey has conveniently summarized the content of some of these other lists:

> Derek Parfit has proposed a list that includes as items moral goodness, rational activity, the development of one's abilities, having children and being a good parent, knowledge, and the awareness of true beauty . . . .James Griffin's list includes accomplishment, the components of human existence (autonomy, capability and liberty), understanding, enjoyment, and deep personal relationships . . . .John Finnis, finally, has proposed a list that includes life, knowledge, play, aesthetic experience, friendship, practical reasonableness, and religion . . . .(Brey 2012, 19)

Objective list theories have also been criticized. One criticism is that they appear dogmatic and paternalistic, declaring certain goods to be constitutive of well-being, whether or not an individual recognizes them as such, thus failing to recognize individual differences. It has also been observed that some items on the list are incomparable. How would we compare a life with little knowledge but great friendships and family relationships to a life with great knowledge but little in the way of friendships and family relationships? (Brey 2012) Furthermore, since the lists are not entirely identical, how can one be sure just which items should be on the list? Nevertheless, the objective list approach offers great promise, especially if desire-fulfillment theories also point toward objective list theories.

Now we can turn to the discussion of well-being in psychology, where the two dominant approaches to the meaning of well-being are the hedonic and eudemonic (Brey 2012, 23). Hedonic psychologists have included the pleasures of the mind as well as the body, and concluded that well-being is primarily subjective well-being (SWB), involving both an affective component (moods and emotions) and a cognitive component (how people perceive their lives in relation to their notion of the ideal life). SWB has been the primary index of well-being in most empirical research, and it is usually measured with a satisfaction-with-life survey that asks the following question: "On a scale of 1–10, how satisfied are you with your life taken as a whole?" This question has been asked to thousands of subjects all over the world, and it yields easily quantifiable results. In agreement with Easterlin, Ryan and Deci have shown that, although people in richer nations are happier than people in poorer nations, the increases in wealth within developed nations have not, in recent decades, been associated with an increase in subjective well-being. Furthermore, in wealthy nations, greater personal wealth shows only a small correlation with happiness. In fact, people who have a strong desire for wealth are less happy than those who do not, whereas those who focus on non-material goals are happier (Ryan and Deci 2001).

Nevertheless, the SWB approach suffers from some serious inadequacies. First, the answer to the life-satisfaction question depends to a considerable extent on how subjects feel at the moment they are asked the question. According to Martin Seligman, one of the founders of contemporary psychology of well-being, "Averaged over many people, the mood you are in determines more than 70 percent of how much life satisfaction you report and how well you *judge* your life to be going at that moment determines less than 30 percent" (Seligman 2011, 13). Introverts and others with low positive affect rank low on the scale, because they are less apt to be in a cheery mood when asked the question. Second, the hedonistic approach does not exhaust activities and states that people choose for their own sake. It does not, that is, include all of the elements that should be included in the full conception of a well-lived life. While we might not want to call Lincoln or van Gogh or Wittgenstein "happy" in the hedonic sense, for example, they probably led deeply meaningful and worthwhile lives. Finally, SWB may not always be a reliable indicator of a healthy living (Ryan and Deci 2001). One could be happy as a drug addict or while engaged in other activities that might not lead to a long and productive life.

These and other problems have lead Seligman to move from the hedonistic approach, embraced in his earlier work, to the eudemonic account of well-being (Seligman 2011). Tracing its lineage to Aristotle, the eudemonic approach understands well-being as concerned with self-realization, and, as Aristotle observed, eudemonic well-being is something that must be gauged over a lifetime, rather than a short span of time (Aristotle 1962, 17–18). Seligman points out that while happiness is a "real thing," a phenomenologically experienced state, well-being is a construct, which has "elements." He identifies five elements of well-being or "flourishing," none of which defines it but each of which contributes to it. The five elements are positive emotion, engagement or enjoyment of activities in which one can be completely absorbed, meaning or belonging to and serving something larger than oneself, accomplishment in projects or work, and positive relationships (Seligman 2011, 16–20).

European scholars have developed a similar concept of flourishing, with associated survey questions (So and Huppert 2009). They found that flourishing was associated with higher educational levels, higher income, and being married. In countries in the European Union, where the surveys were conducted, the highest level of flourishing was in Denmark and the lowest level in Russia. The United Kingdom had about half the rate of flourishing as Denmark (Seligman 2011). In another similar account, Ryff and Keys identify six factors in eudemonic well-being which show considerable similarity to Seligman's list: self-acceptance, positive relations with others, autonomy, environmental mastery, purpose in life, and personal growth (Ryff and Keys 1995). While the hedonic account of well-being is still dominant in psychology, the eudemonic account may be more conceptually adequate.

The similarities between the objective list account of well-being in philosophy and the eudemonic account in psychology are obvious. Both suggest a set of goods which comprise well-being. These include the goods of human relationships and family, purpose and meaning in life, cultivation of one's abilities and rational faculties, achievements, and (in several lists), aesthetic appreciation. However, the specificity of these accounts poses a problem. Should engineers or any other group presume to decide what constitutes the good life? To what extent is it possible to move away from the paternalistic prescriptivism implicit in the objective list approach?

7.4 The Capability Approach and the Conditions of Well-Being

In the "Foreword" to *A Century of Innovation,* mentioned earlier, Neil Armstrong suggests one possible way to move away from excessive paternalistic prescriptivism. Referring to his observation that "we each may have our own definition of the term 'quality of life,'" Armstrong remarks that "most of us would probably acknowledge that certain living conditions are essential to a preferred quality in our own lives" (Consable and Somerville 2003, vi). Armstrong apparently thinks of the twenty greatest engineering achievements of the twentieth century as conditions of well-being, however one might define well-being. This position, as has been indicated, is unsatisfactory. A more abstract and time-independent account of the conditions of well-being is needed.

An alternative account of the conditions of well-being is supplied by the Capability Approach (CA). Pioneered originally by Nobel laureate Amartya Sen and philosopher Martha Nussbaum (Sen 1999; Nussbaum 2000), the CA was originally intended to guide economic development in less-industrialized societies, but the essential ideas of the CA are applicable for developed nations as well (Robyens 2011). For Sen, poverty, in its most essential meaning is not the absence of wealth or material and technological advantages, but the deprivation of opportunities or capabilities to achieve well-being. "Capabilities" are freedoms to pursue well-being, which Sen calls "well-being freedom" (Sen 1992, 40). He further describes capabilities as

> ...the various combinations of functioning (beings and doings) that a person can achieve...a set of vectors of functionings, reflecting the person's freedom to lead one type of life or another...to choose from possible livings....(Sen 1992, 40)

We can conclude that increasing capabilities, as the necessary conditions for achieving well-being, is the goal of technological development.

What are the capabilities that are the necessary conditions of well-being? Sen has declined to propose "one pre-determined canonical list of capabilities, chosen by theorists without any general social discussion or public reasoning" (Sen 2005, 158). Nussbaum has not been as reluctant, identifying a set of basic capabilities. The capabilities she identifies are needed in order for a human life to be "not so impoverished that it is not worthy of the dignity of a human being"(Nussbaum 2000, 72). For purposes of engineering, Nussbaum's list has distinct advantages over the lists proposed by philosophers and psychologists, not only because it is more comprehensive, but also because the capabilities do not, strictly speaking, identify the content of well-being, but rather the conditions of well-being, however it may be defined. Nussbaum has identified a list of "Central Human Functional Capabilities" (Nussbaum 2000, 78–80). In the following version of Nussbaum's list, I have rearranged the capabilities she identifies under four categories and supplied comments on their relationship to technology. This arrangement brings us a step closer to a version of capability theory useful for engineers.

Group I: Physical Capabilities

- Living a normal length of life.
- Having clean water, food, and shelter.
- Engaging in recreational activity.

The links between physical health and well-being are complex (Diener and Seligman 2004). A sense of well-being leads to a longer and healthier life, and a longer and healthier life leads to well-being (David et al. Introduction 2013). Engineering contributes to physical well-being in many ways. Of the twenty greatest engineering achievements, water supply and distribution, agricultural mechanization, air conditioning and refrigeration, imaging, household appliances, and health technologies are especially important.

Group II: Capabilities for Human Relationships

- Having love and attachments to things and other people.
- Being treated with respect and dignity.

Aristotle emphasized the importance of friendship for happiness. Friends value and respect each other, in part because of the mutual benefit that the relationship affords (Pangle 2003). Contemporary research has confirmed the strong link between relatedness and well-being (Ryan and Deci 2001). Some evidence suggests that the quality of friendships is in general more important than the quantity. Research has focused on the relationships of older people, but this conclusion is probably applicable to all ages. Technology can contribute to the development of quality relationships in many ways, but one example is social networking on the Internet. Individuals with like interests can connect, often over large distances. Many romantic relationships, often leading to marriage, have started in this manner. The Internet has also enabled individuals with similar physical or mental disabilities or simply those suffering from loneliness and isolation to form supportive bonds. Probably may gay youths have been kept from suicide because of Internet connections. Some of the engineering achievements that contribute to, and even enlarge, the capabilities for human relationships are electrification, electronics, and radio and television.

Group III: Social/Political Capabilities

- Moving about freely and safely.
- Using one's senses and imagination and having free expression.
- Being able to participate in the political process, preserve material goods, and hold property.

Free movement and expression, political participation, and property rights are associated with democratic institutions, so it should come as no surprise that the well-being of citizens is generally higher in democratic societies, especially in direct democracies in which citizens can participate in decision-making (Frey and Stutzer 2000). The use of cell phones and other modes of technology has apparently facilitated recent social change in the Middle East. Other engineering achievements that have contributed to communication and the ability to move freely are computers, telephony, the automobile and airplane, highways, spacecraft, petroleum and petroleum technologies, and nuclear technologies.

Group IV: Capabilities for Self-Transcendence and Meaning

- Being able to form a conception of the good life and to plan one's life
- Living with concern for and in relation to nature.

Identification with something larger than oneself has been recognized by most spiritual traditions as an important aspect of having a meaningful sense of one's life. Aristotle believed that a meaningful life was one lived in community. The spiritual traditions of China and in many other cultures hold that happiness is obtained by being connected to a larger pattern or order beyond human society. One mode of connection is by way of nature. People whose lifestyle is supportive of the natural world tend to have higher levels of SWB (Brown and Kasser 2005). Engagement with the natural world tends to produce lower levels of stress and aggression and higher levels of well-being (Newton 2007). Technology can contribute to the access to the natural world, whether it be through transportation to places of natural beauty and interest or the development of high-tech running shoes or hiking equipment.

Two observations about these four categories are relevant. First, these four categories bear a remarkable similarity to the accounts of well-being proposed by philosophers and psychologists, and especially to the summary list proposed earlier. Category IV (meaning) and Category II (human relationships) have their close parallels in the lists in philosophy and psychology. The importance of self-cultivation and personal achievement may in a general way correspond to Category III. Category I comprises the physical conditions of well-being.

Second, these categories correspond roughly to the capabilities that are of central importance in both developing and developed nations. Evidence already cited, however, suggests that the relationship of technological development to well-being may be different for technologically developing and developed societies. Technological development seems to promote well-being up to a certain threshold, but, after that, becomes less efficacious in achieving this goal. Translated into capability theory, this means that the connection between economic and technological progress and increases in capabilities for well-being may be different in developing and developed societies.

This distinction between capabilities relevant to developed and developing nations is by no means exact, but it is nevertheless useful in considering the relation of technology to capabilities and in sorting out those capabilities in terms of those which are more immediately relevant to nations that have a relatively low level of technological development and nations that have a relatively high level of technological development. The capabilities most relevant to developing nations are those most directly relevant to physical survival. Thus Category I has special relevance to nations with relatively low levels of technological development, since Category I comprises those capabilities that are closely related to the physical conditions of life. Categories II, III and IV have special relevance to nations with relatively high levels of technological development, where basic physical needs that are important for survival have been met. In these societies, concern is more likely to focus on democratic freedoms, having satisfying human relationships, and finding the basis for a meaningful and purposeful life. The negative effects of the technologies included in Category I, however, are also relevant to developed nations.

7.5 Designing for Well-Being

The defining professional activity of engineers is design, and the fundamental thesis of this paper is that engineers should give more serious consideration to the impact of their designs on human well-being, both in developing and developed societies. The time may be approaching when "well-being impact" will be a standard design constraint, much as environmental impact is today. If the argument of this essay is valid, increasing capability for well-being should be an explicit goal of engineering design.

Lest the suggestion that well-being (or capability for well-being) should become a design constraint be considered too far-fetched, it is worthwhile to note that well-being is already emerging as a constraint in public policy. The King of the tiny Asian country of Bhutan endorsed his father's words that "Gross National Happiness (GNH) is more important than Gross National Product (GNP) (Thimphu 2013). The governments of the UK, France, Canada, and Australia are initiating movements to measure well-being on a national scale (David et al. 2013. Introduction.). For governments, coming up with quantitative measures of well-being is a major challenge, especially since some research suggests that the hedonic and eudemonic approaches may diverge in ways that are difficult to reconcile (Ryan and Deci 2001). Some type of quantification may be necessary if well-being is to become a design constraint in engineering.

It appears, however, that there is a fair amount of consensus on the factors that are important in happiness, which is at least a close kin of well-being. According to a poll on the factors that influence happiness conducted by the British Broadcasting Corporation, partner/spouse and family relationships account for 47 % of the total and health accounts for another 24 %. The principal remaining factors are work fulfillment (2 %), community and friends (5 %), religion/spiritual life (6 %), money and financial situation (7 %), and a nice place to live (8 %) (Mulgan 2013. Introduction.). Already, these numbers suggest the areas which should be of special interest to engineers concerned with designing to promote capabilities for well-being.

Designing for well-being, however requires a critical attitude toward technology itself. The account of the twenty greatest engineering achievements of the twentieth century appears to reflect an uncritical technological optimism. Interpreted in terms of capability theory, the claim of technological optimism is that technology always and everywhere enhances and never detracts from human capabilities for well-being. By contrast, some writers, especially in the humanities and social sciences, have called attention to the downside of technology, that it can sometimes detract from the opportunity to lead a good and fulfilling life, especially with regard to interpersonal relationships, finding meaning and overall purpose in life, and our relation to the natural world. Developing a critical attitude toward technology requires an understanding of technology's effects on human life, both for good and ill. It is worth noting in this regard that the leadership of the engineering profession has itself recognized the need for engineering students to understand the social effects of technology, although the motive may or may not be to promote a critical attitude toward technology. In 2000, the Accreditation Board for Engineering and Technology (ABET) officially recognized the importance of social awareness in its criteria for engineering education. Criterion 8 of ABET's guidelines for engineering education requires engineering students to have "the broad education necessary to understand the impact of engineering solutions in a global and social context" and Criterion 10 requires students to have "a knowledge of contemporary issues" (onlineethics.org/Education/instructessays/herkert2.aspx).

In terms of the CA, the criterion for evaluating a particular technology is *whether the technology enhances or diminishes the capabilities of those who use it for achieving a life they have reason to value.* While applying this criterion can be enormously challenging, consider the following sketch of a procedure that a designer might use in applying this criterion.

First, the designer must identify the capabilities that might be affected by his or her design. We can call these the "relevant capabilities. " In designs for use in developing nations the capabilities in Category I are likely to be more crucial; in designs for use in developed nations, the capabilities in Categories II, III, and IV are likely to be more crucial.

Second, the designer must determine whether and how the relevant capabilities are augmented or diminished by the design. This determination is complex, because a design may augment one capability and diminish another or both augment and diminish the same capability. Or, a design may augment or diminish capabilities for one group of individuals and have a very different effect on another group.

Third, the designer must determine whether the effects of the design on the relevant capabilities are acceptable. We can call this the determination of the "well-being acceptability" of the design. This determination must be based not only on whether the net augmentation of capabilities is greater than the diminishment, but whether some capabilities are diminished below a minimal acceptable level.

Fourth, if the design is determined to have an unacceptable effect on capabilities, a new design which has a higher well-being acceptability should be attempted.

7.6 Designing for Developing and Developed Nations

Now let us look at the issues that might arise in designing for developing and developed nations.

First, consider developing nations. Much of the literature on aspirational ethics focuses on the promotion of well-being, or the capabilities for well-being, in developing nations. The need for this focus is obvious. It is estimated that 20 % of the human population lacks clean water, that 40 % does not have adequate sanitation, and that 20 % is in need of adequate housing (Amadei 2004). In the face of these numbers, several scholars and engineering groups have undertaken to advocate technological development in less developed parts of the world. In 2008 the NAE hosted a conference on "Engineering, Social Justice, and Sustainable Communities" (National Academy of Engineering 2010). Scholar Carl Mitcham has come up with the term "humanitarian engineering," which is defined as "design under constraints to directly improve the well-being of under-served populations"(Mitcham and Munoz 2010, xi). He has been instrumental in establishing a program in humanitarian engineering at the Colorado School of Mines. Bernard Amadei, founder of Engineers Without Borders-USA, holds that "the engineering profession must revisit its mindset and adopt a new mission statement—to contribute to the building of a more sustainable, stable, and equitable world"(Amadei 2004, 25). In the same paper he lists two aims of Engineers Without Borders—USA: "(1) to help disadvantaged communities improve their quality of life through implementation of environmentally and economically sustainable engineering projects, and (2) to develop internationally responsible engineering students" (Amadei 2004, 28).

Amadei has been instrumental in establishing at the University of Colorado at Boulder a program somewhat similar to Mitcham's. Doubtless neither Mitcham nor Amadei want to maintain that every engineer has an obligation to travel to developing countries to promote technological development, but they do appear to hold that the engineering profession should support initiatives in this direction and that the engineering curriculum should provide a place for educating engineers interested in addressing the problems of developing nations. According to Amadei, such a curriculum should "include water provisioning and purification, sanitation, power production, shelter, site planning, infrastructure, food production and distribution, and communication, among many others" (Amadei 2004, 25).

The critical attitude is important in promoting capabilities for well-being in developing nations because not all technology achieves this goal and technological development can even diminish capabilities for well-being. As Lucena et al. have pointed out, design for industry is not necessarily appropriate for developing societies. Designs and off-the-shelf parts appropriate for the US market may not be appropriate for developing societies. Costs may be entirely inappropriate for developing societies. A product costing $800 is not affordable in a country with an annual GDP of $1600. A design appropriate for a Senegalese community may not be appropriate for communities in India. Furthermore, members of a community in a developing society may not accept a pilot design demonstrated to them when their voices have been excluded in the earlier phases of the design process (Lucena et al. 2010). In engineering for developing communities, some understanding of the communities themselves is required. One must understand the relationships among the members of the community and especially the power structures, the relationship of the community to "place," and the differentiation of gender roles in the community. Best practices in one community may not be best practices in another (Amadei and Wallace 2009).

Here is an example of two projects, one being unattractive to villagers in the Western African nation of Ghana, and another being much more attractive. Engineering students developed an ethanol cooking stove which eliminated the smoke in village dwellings that the students would have found very offensive. However, the villagers had become accustomed to the smoke and accepted it as a part of their way of life. Furthermore, ethanol was relatively expensive to produce or buy, while firewood was free. On the other hand an electric light powered by a low cost thermoelectric generator was sufficient to light up a small room so that a book could be read at night. Furthermore, the electricity could be used to recharge cell phones, which were highly prized by the villagers. While many of the homes had electricity, it was unreliable and expensive. So while one product was not accepted by the villagers, the other one was a rousing success (Lucena et al. 2010).

While preventive and prohibitive ethics are best formulated in rules, aspirational ethics is best formulated in a set of virtues. The development of such virtues is therefore of crucial importance in motivating aspirational ethics. Two virtues that are especially important in aspirational ethics for developing societies are empathy and compassion. Empathy manifests itself in many ways, but one of the most important is the ability to listen to those for whom the technologies are intended. The needs, attitudes, social relationships, and values of individuals in developing societies must be understood and appreciated. Listening is a capacity which is developed only in practice and is ordinarily not taught in engineering schools, or at least not in engineering courses. Compassion likewise is not a virtue ordinarily taught in engineering schools. While it is common for physicians to embrace compassion as a professional virtue, engineers may be less comfortable with it. Nevertheless, one engineering student has advocated in my presence the importance of a "caring heart," especially with regard to work with developing societies. Amadei has expressed a similar concern for the disadvantaged (Amadei and Wallace 2009).

Now we can turn toward designing for developed nations, keeping in mind the critical attitude toward technology described earlier. One way to appreciate the importance of the critical attitude, especially as it applies to developed nations, is to take a second look at the four categories of capabilities listed above and think about the ways technology can endanger well-being as well as promote it. With regard to the first category of physical capabilities, for example, technology can pollute the air, water and food supply with toxic chemicals and promote a sedentary and unhealthy lifestyle. With regard to the second category of capabilities for human relationships, technology can impede effective human communication, as well as enhance it. Shannon Vallor has argued that social networking can inhibit the development of virtues necessary for quality social relationships (Vallor 2012). One recent study has confirmed that online social interactions lead to offline loneliness and depression (Kang 2007). With regard to the third category of social/political capabilities, technology has provided the basis for massive invasions of privacy, data mining and consequent invasive marketing, and other types of breaches of personal space. With regard to the fourth category of capabilities for self-transcendence and meaning, the research data is more incomplete. To the extent that technology contributes to a consumer culture that stimulates desire for wealth and the possession of commodities, it may detract from the ability to find the deeper sources of happiness (Ivanhoe 2013). Some argue that technology is also an impediment to the spiritual life. In an extended inquiry into the relationship of technology to spirituality, Albert Borgmann argues that science and technology have diminished the experience of community and ritual that are important in finding a connection with something larger than oneself. Instead, we settle for isolated and non-participatory experiences. Technology characteristically increases efficiency and comfort, but often at the cost of connection with a larger community or the natural world (Borgmann 2003). Biblical scholar Dale Alison has argued that "the more we have moved indoors, the less some of us have been inclined to believe." With increasing mastery of nature, we have lost the sense of wonder that is crucial for openness to transcendence (Alison 2006).

Even this brief discussion suggests several virtues that are especially important for the aspirational attitude with respect to developed societies: concern for the environment, sensitivity to the effects of technology, especially on human society and relationships, and creativity. With the possible exception of concern for the environment, these virtues may or may not be taught or encouraged in engineering education. These virtues are most likely to be acquired from courses in philosophy of technology, the history of technology, Science and Technology Studies, biology, and environmental ethics. Again, the importance of humanities and non-engineering courses for engineers is evident.

References

ACM. (1992). ACM code of ethics and professional conduct. Association for computing machinery. http://www.acm.org/about/code-of-ethics. Accessed 25 Oct 2013.

Adler, M. D. (2012). *Well-being and fair distribution*. New York: Oxford University Press.

Alison, D. (2006). *The luminous dusk: Finding god in the deep still places*. Grand Rapids: Eerdmans.

Amadei, B. (2004). Engineering for the developing world. *The Bridge, 32*(2), 24–31.

Amadei, B., & Wallace, W. A. (2009). Engineering for humanitarian development. *IEEE Technology and Society Magazine, 28*(4), 6–15.

Angeles, L. (2011). A closer look at the Easterlin Paradox. *Journal of Socio-Economics, 40*(1), 67–73. doi:10.1016/j.socec.2010.06.017.

Aristotle. (1962). *Nicomachean ethics*, (trans: Martin Oswald). Indianapolis: Bobbs-Merrill.

ASCE. (2008). Ethics, guidelines for professional conduct for civil engineers. American Society of Civil Engineers. http://www.asce.org/uploadedFiles/Ethics_-_New/ethics_guidelines010308v2.pdf. Accessed 25 Oct 2013.

Board of Ethical Review. (1982). Case 82.5. Whistleblowing. http://www.Nspe.Org/resources/ethics/ethics-resources/board-of-ethical-review-cases.

Borgmann, A. (2003). *Power failure: Christianity in the culture of technology*. Grand Rapids: Brazos Press.

Brandt, RB. (1998 [1979]) *A theory of the good and the right*. Amherst: Prometheus Books.

Brey, P. (2012). Well-being in philosophy, psychology, and economics. In P. Barey, A. Briggle, & E. Spence (Eds.), *The good life in a technological age* (pp. 15–34). New York: Routledge.

Brown, K. W., & Kasser, T. (2005). Are psychological and ecological well-being compatible? The role of values, mindfulness, and lifestyle. *Soc Ind Res, 74*, 349–368.

Constable, G., & Somerville, B. (2003). *A century of innovation*. Washington, DC: Joseph Henry Press.

David, S. A., Boniwell, I., & Conley Ayres, A. (2013). *The oxford handbook of happiness*. Oxford: Oxford University Press.

Diener, E., & Seligman, M. E. P. (2004). Beyond money: Toward an economy of well-being. *Psychological Science in the Public Interest, 5*(1), 1–31.

Easterlin, R. A. (1974). Does economic growth improve the human lot? Some empirical evidence. In R. Davis & M. Reder (Eds.), *Nations and households in economic growth: Essays in honor of Moses Abramovitz* (pp. 89–125). New York: Academic.

Easterlin, R. A. (1995). Will raising of all increase the happiness of all? *Journal of Economic Behavior and Organization, 27*(1), 35–47.

Easterlin, R. A. (2003). Explaining happiness. Proceedings of the National Academy of Sciences of the United States of America, 100(19), 11176–11183.

Frey, B. S., & Stutzer, A. (2000). Happiness, economy, and institutions. *The Economic Journal, 110*, 918–938.

IEEE. (2006). IEEE code of ethics. The Institute of Electrical and Electronics Engineers. http://www.sfjohnson.com/acad/ethics/FourCodesOfEthics.pdf. Accessed 25 Oct 2013.

Ivanhoe, P. J. (2013). Happiness in early Chinese thought. In S. A. David, I. Boniwell, & A. Conley Ayres (Eds.), *The Oxford handbook of happiness* (pp. 263–278). Oxford: Oxford University Press.

Kang, S. (2007). Disembodiment in online social interaction: Impact of online chat on social support and psychosocial well-being. *Cyber Psychology & Behavior, 10*(3), 475–477.

Lucena, J., Schneider, J., & Laydens, J. A. (2010). *Engineering and sustainable community development*. San Rafael: Morgan and Claypool.

Mitcham, C., & Munoz, D. (2010). *Humanitarian engineering*. San Rafael: Morgan and Claypool.

Mulgan, G. (2013). Well-being and public policy. In S. David, I. Boniwell, & A. Conley Ayres (Eds.), *The Oxford handbook of happiness* (pp. 517–532). Oxford: Oxford University Press.

National Academy of Engineering. (2010). *Engineering, social justice and sustainable community development*. Washington, DC: The National Academies Press.

Newton, J. (2007). *Wellbeing and the natural environment: A brief overview o the evidence*. London: Department for Environment, Food, and Rural Affairs.

Nozick, R. (1974). *Anarchy, State, and Utopia*. New York: Basic Books.

NSPE. (2007). Code of ethics for engineers. National Society of Professional Engineers. http://www.nspe.org/resources/pdfs/Ethics/CodeofEthics/Code-2007-July.pdf. Accessed 25 Oct 2013.

Nussbaum, M. (2000). *Women and human development: The capabilities approach*. Cambridge, UK: Cambridge University Press.

Pangle, L. S. (2003). *Aristotle and the philosophy of friendship*. Cambridge, UK: Cambridge University Press.

Robyens, I. (2011). The capability approach. In Edward N. Zalta (Ed.) *The Stanford encyclopedia of philosophy*. http://plato.stanford.edu/entries/capability-approach/

Ryan, R. M., & Deci, E. L. (2001). On happiness and human potentials: A review of research on hedonic and eudemonic well-being. *Annual Review of Psychology, 52*, 141–166.

Ryff, C. D., & Keys, C. L. M. (1995). The structure of psychological well-being revisited. *Journal of Personality and Social Psychology, 69*, 719–727.

Seligman, M. (2011). *Flourish: A visionary new understanding of happiness and well-being*. New York: Free Press.

Sen, A. (1992). *Inequality re-examined*. Oxford: Oxford University Press.

Sen, A. (1999). *Development as freedom*. New York: Knopf.

Sen, A. (2005). Human rights and capabilities. *Journal of Human Development, 6*(2), 151–166.

So, T., & Huppert, F. (2009). What percentage of people in Europe are flourishing and what characterizes them? www.isqols2009.istitutodeglinnocenti.it/Content_en/Huppert.pdf. Accessed 25 Oct 2013.

Thimphu, T. (2013). Foreword. In A. David Susan, I. Boniwell, & A. Conley Ayres (Eds.), *The Oxford handbook of happiness* (pp. xvii–xviii). Oxford: Oxford University Press.

Vallor, S. (2012). New social media and the virtues. In P. Brey, A. Briggle, & S. Edward (Eds.), *The good life in a technological age* (pp. 193–202). New York: Routledge.

Wulf, W. A. (2000). Great achievements and grand challenges, revised version of a lecture given on October 22 at the 2000 annual meeting of the National Academy of Engineering, p. 1. Used with permission.

Chapter 8
Towards an Ethics of Technology and Human Development

Ilse Oosterlaken

Abstract One of the societal challenges that engineering in a global world faces is that of making technology work in the context of developing countries and poverty reduction, to make it truly contribute to human development. This makes the relatively young field of development ethics potentially highly relevant to engineering, but unfortunately it has so far hardly addressed technology. To make its application to technology more than superficial, it is important to thoroughly explore its connections to engineering ethics, to ethics of technology, and even philosophy of technology more broadly. This claim is illustrated with the so-called 'capability approach', which is nowadays very popular within development ethics and which attaches central moral importance to individual human capabilities. The chapter discusses how insights from philosophy and ethics of technology are useful, among others, to better conceptualize the relation between technical artifacts and valuable human capabilities. In this way the chapter makes a small theoretical contribution towards an endeavor to create an ethics of 'technology and human development.'

Keywords Development ethics • Ethics of technology • Capability approach • Engineering ethics • Design • Socio-technical systems

8.1 Introduction

One of the challenges that engineering in a globalized world faces is that of making technology work in a developing countries context. Some engineering programs have explicitly taken up this challenge. A prominent example is the "Humanitarian Engineering" program at the Colorado School of Mines, which started in 2003. It "seeks to prepare engineering students for careers that will benefit the underserved

I. Oosterlaken (✉)
VU University Amsterdam, Amsterdam, The Netherlands
e-mail: E.T.Oosterlaken@vu.nl

© Springer International Publishing Switzerland 2015
C. Murphy et al. (eds.), *Engineering Ethics for a Globalized World*, Philosophy
of Engineering and Technology 22, DOI 10.1007/978-3-319-18260-5_8

international community" (Moskal et al. 2008). On its website[1] it is explained that "in the past, engineers may have asked, 'How do I generate electricity most efficiently?' The humanitarian engineer asks, 'How can I help to reduce poverty?'" Other, related reasons why some universities have started paying attention to 'development' within engineering curricula are that it is seen as an integral part of sustainability education (Boni and Perez-Foguet 2008), that it is one way to encourage a spirit of voluntary or social service amongst engineers (Passino 2009), and that it may help to cultivate engineers' humanity and foster their cosmopolitan abilities (Boni et al. 2012).

Another type of motive can be found at my own university, Delft University of Technology. Its Faculty of Industrial Design Engineering has in the past decade sent dozens of students to the Global South for design projects. Both staff and students are inspired by the work of business scholars Prahalad and Hart, who identified a huge market and good business opportunities at the so-called 'Base of the Pyramid' (BoP – referring to people living on less than a \$1/day), with the promise that profit and poverty reduction can go hand-in-hand. Design projects are often done in collaboration with companies (Kandachar et al. 2011). The same starting point is present in the minor "Entrepreneurship & Development" of the neighboring Faculty of Technology, Policy and Management, a minor which is quite popular with Delft's engineering students.

In short, in our globalized world there are both economic and humanitarian reasons to prepare engineers for working in the context of developing countries, and some even speak of a "trend" in engineering education to do so (Vandersteen et al. 2009). One important question then becomes what engineers need to learn in order to be successful in the Global South. Robbins (2007) concludes from interviews with over 30 'Northern' engineers with relevant experience that working in this context requires "reflexive engineers." Such engineers, he says, differ in several ways from "traditional engineers." He distinguishes nine, somewhat related, dimensions. One of these is that traditional engineers would take "designs" as their "conceptual starting point", whereas reflexive engineers, equipped to work in developing countries, would focus on "socio-technical systems" – I will get back to this later on in the chapter. Another difference is that the "view of development" of traditional engineers would be "technology driven", whereas reflexive engineers would take a "livelihoods based" perspective. Although I support moving away from a fixation on technology, by instead asking to which development goals technology should contribute, I would like to note that the livelihoods approach is a quite specific take on development (see e.g. Zoomers 2008). Hence stimulating real reflexivity requires placing this, in turn, into broader critical debates about the ends and means of development. Such debates are taking place within development studies, but increasing also in a specialized sub-discipline called development ethics.

In this chapter I won't say much in particular about what sort of education the 'humanitarian engineer' requires, but my basic assumption is that development

---

[1]http://humanitarian.mines.edu/, accessed September 16th 2012.

ethics should be an integral part of any such program[2] – for which purpose standard textbooks on ethics for engineering students do not suffice.[3] To make development ethics relevant to engineers, however, and to make its application to technology more than superficial, it is important to thoroughly explore its connections to engineering ethics, to ethics of technology, and even philosophy of technology more broadly.[4] The result should arguably be an ethics of 'technology and human development' that acknowledges that both development and technology are value-laden, and that fruitful ethical reflection on the topic requires – contra Robbins' statement – paying attention to both engineering design and socio-technical systems.

This chapter makes a theoretical contribution towards such an endeavor by firstly introducing development ethics (Sect. 8.2). I will then argue that the time seems ripe for making more explicit connections between development ethics and ethics of technology (Sect. 8.3). To illustrate how the two fields may benefit from each other, and how connections may be forged, I will discuss the application of the so-called 'capability approach' to technology. This approach has become very popular within development ethics, but only recently have scholars started to use it as a normative lens for looking at technology (Sect. 8.4). The next section will theorize the technology – human capabilities relationship in some more depth, arguing that both the details of design and the socio-technical embedding of technical

---

[2]Of course, not all technology projects and engineering efforts in the global South explicitly aim at contributing to development. It may therefore be useful to distinguish between 'technology for development' and merely 'technology in developing countries' – as Brown and Grant (2010) also propose to do for ICT for Development (ICT4D) research. One might argue that it makes sense for engineering students involved in the former to get introduced to development ethics, but that this not needed for the latter group. Yet, so Robbins and Crow (2007, p. 77) point out, even if not part of "intentional development", it is still the case that "design work for a corporation constructing a hotel, or other commercial development, constitutes work within *immanent development*. In both roles, engineers may also contribute to the accretion or generation of a vision of development." Their view, which I would like to support, is that "reflexivity involves an awareness of these different locations (immanent, intentional, vision-making) in trajectories of social change, and their relation to each other." Hence, in a globalized world development ethics may also offer a useful expansion of the ethics training of a much broader group of engineering students, not just the ones planning to work in "intentional development."

[3]Whitbeck (1998) just mentions an engineer building a water pump in Latin-America in the epilogue, as one example of finding meaningful work in the engineering profession. Van de Poel and Royakkers (2011) do a little bit better by discussing the Shell/Nigeria/Ogoni case, codes of conduct for multinationals, and – very briefly – some general ethical principles as discussed by Luegenbiehl (2010) and Harris (1998). The developing country context is most extensively being discussed in the textbook by Harris et al. (2000), which includes a chapter "International Engineering Professionalism" that discusses e.g. common conditions in the South, codes of conduct in international context, human rights, and the problems of paternalism and exploitation. This certainly suffices for a general introduction course in ethics, but the 'humanitarian engineer' would need more.

[4]This should, among others, lead to acknowledging that technology affects human lives in many ways beyond extending or destroying people's livelihoods – and that many of these impacts will also be relevant from an ethical perspective.

artifacts are relevant factors in the expansion of human capabilities. A case study of podcasting devices in Zimbabwe will be used as a brief illustration (Sect. 8.5). I will end with some conclusions (Sect. 8.6).

8.2   A Brief Introduction to Development Ethics

Development ethics studies the ends and means of development, and the responsibility of people and institutions for this process, from a distinct moral perspective. This can be either at local, national, regional or global level. A central question in development ethics of course concerns the meaning of 'development' itself, and the normative evaluation of alternative understandings, models and paths. A basic requirement for such critical thinking is abandoning the deterministic idea that societies can only develop in one universal way, which was popular in post-WWII development thinking and exemplified in e.g. Rostow's famous five *Stages of Economic Growth*. Equating development to economic growth itself has, of course, also received much criticism over time – and some post-colonial activists and social critics, attacking this orthodox economic development thinking by proposing other development goals, could be considered as engaging in development ethics (Crocker 2008).

Even if we would all agree that *good* development is ultimately about making people's lives better, there is still a lot of disagreement possible on how this should be understood, and which values then come into play. Surely well-being is a prominent value, but even that can be interpreted in a myriad of ways. One subjective interpretation is e.g. that well-being equals happiness, so that development should promote whatever makes people happy. One objective interpretation is e.g. that promoting well-being means assuring that some universal human basic needs are met for everyone. Development ethics critically scrutinizes such interpretations, and investigates their moral justifications and implications.

Another objective view on well-being, one which has become very influential in development ethics, is the 'capability approach' founded by economist Amartya Sen (1999) and philosopher Martha Nussbaum (2000). It conceptualizes well-being in terms of individual human capabilities – i.e. what a person is effectively able to do and be. Examples are the capability to be healthy, to get an education, to be part of a community, or to travel. Development is then defined as a process of expanding people's set of such capabilities. One thing that makes this approach normative, is that it is first and foremost interested in *valuable* capabilities, those that are a constitutive part of flourishing human lives. Nussbaum has for example created a list of 10 such capabilities, and argues that justice requires that each and every person is brought up to at least a threshold level of these capabilities. Another value central in the capability approach is human agency. It considers the poor not as patients to be helped, but as active agents in changing their lives and their society. This is not merely a descriptive claim, but also a normative claim: people *should* be treated as full agents. An extension of people's capability set means an extension of the degree to which they are able to act as an agent.

In defending this approach, capability scholars of course also criticize alternative views on development, mainly those focused too much on either subjective well-being, or on merely means to well-being, such as income and other resources. The problem with the latter is that these do not always, everywhere and for everyone translate into human capabilities, due to so-called personal, social and environmental "conversion factors." A bicycle does, for example, not convert into a capability to travel for people with certain disabilities, women in Iran, or people living in deserts without paved roads. The capability approach draws attention to the fact of human diversity, or the pervasiveness of such differences. Human diversity is also acknowledged in a different way, namely that people have different views of the good life. Making sure that people have an extensive set of human capabilities means empowering people to realize the specific life they have reason to value. The philosophical work on the capability approach shows that development ethicists pay a lot of attention to the overall normative foundations of development, and their justification. The capability approach has also led to a lot of empirical work, for example using the conceptual framework that it provides to assess the development level of countries, or to evaluate development projects. It could thus be taken to illustrate the interdisciplinary character of development ethics.

One key theme in development ethics is furthermore the question of a fair distribution of the costs and benefits of development initiatives, both between groups in a society and between present and future generations. This is an ethical question that is salient, for example, in the many dam-related settlement cases that can be found in developing countries (Gasper 2012). Furthermore, ethical dilemmas arise in both grass roots development initiatives and development policy when value conflicts occur, "between different people who have different values, or have different priorities even where there are shared basic values" (Dower 2008, p. 189). The harmoniously sounding term 'sustainable human development', for example, may mask that concrete development initiatives increasing human well-being may be at the expense of biodiversity and other ecological aspects – and vice versa. And the gender norms of Western development organizations will often clash with those in male-dominated societies. Finally, there are also ethical questions concerning the responsibilities of various actors active within development endeavors, and what would count as virtues for development professionals (Crocker 2008). Many also consider a broad range of questions at an international scale, such as the global justice questions studied in e.g. political philosophy, to be part of development ethics. This includes questions such as: "why ought rich countries and/or rich individuals to give aid to help very poor people and/or countries? What is an acceptable basis for international trade and investment? Should crippling third world debt be cancelled?" (Dower 2008, p. 185). One point of disagreement on scope is whether development ethics should focus on poor countries only, or also address deprivations and socio-economic change in overall affluent countries (Crocker 2008).

Although development ethics draws on much older work from adjacent intellectual traditions and philosophical disciplines, it is still relatively young as a distinguishable scholarly discipline – as indicated by on-going discussions about scope, rationale, boundaries, approach, leading questions and so on (Dower 2008; Gasper 2012). It first became institutionalized in 1984, with the establishment of IDEA, the International Development Ethics Association.[5] Several historical sources, so Crocker (2008) explains, can be identified for development ethics as a separate discipline. The main pioneer in the 1960s/70s was development scholar Denis Goulet, who started reflecting on value issues that he perceived to be present in development practice and policy. Furthermore, in the 1970s Anglo-American philosophers started – stimulated by the writings of Peter Singer and others – discussing if we have any moral duty to provide famine relief. This philosophical engagement with poverty was, however, still quite abstract and limited in several ways. These and other sources, claims Crocker, have since then led to a scholarly field of development ethics that is interdisciplinary yet explicitly normative, empirically informed and context-sensitive. This is nicely illustrated by his chapter on subsequent changes in research and thinking concerning the ethics of hunger and famine relief (Crocker 2008). Prominent thinkers within development ethics, however, would still like to see even more 'bottom-up' or case-study inspired work within development ethics. This should lead to ethical methodologies, insights, principles and considerations that could more directly guide policy-makers and practitioners in making value-conscious decisions (Crocker 2008; Gasper 2012).

8.3 The Gap Between Development Ethics and Ethics of Technology

The past section has given the reader an introduction into development ethics. As I assume that most readers of this book have at least a basic familiarity with ethics of technology, I won't introduce this as extensively. Yet it is interesting to briefly reflect on its history, as some parallels can be drawn with development ethics. For a long time the dominant but implicit view of technology was instrumentalism: technology is a value-neutral means towards human ends. It is then not technology that invites ethical reflection, but merely the actions and values of people using technology. In the early twentieth century philosophers started criticizing technology itself. These classic philosophers of technology saw 'Technology', however, as a monolithic phenomenon and an autonomous force, for example arguing that it was alienating or subjugating everything to the value of efficiency. Their view was also one of technological determinism: we cannot influence the course of technological development, and new technology in turn fully determines how society develops.

Contemporary ethics of technology has been made possible by a combination of three changes as compared to the situation just sketched, and this is where parallels can be drawn with development ethics. Firstly, it is nowadays widely believed that technologies tend to be non-neutral or value-laden and are thus themselves worth

---

[5] www.developmentethics.org

studying from an ethical perspective, although it should be noted that "the value-ladenness of technology can be construed in a host of different ways" (Franssen et al. 2009). Likewise, development ethics has arisen from a realization that the concept and practice of development is thoroughly value-laden. Secondly, since the 1960s the idea of full-blown technological determinism has – under the influence of the new field of Science and Technology Studies (STS), which showed that technology and society continuously co-shape each other – gradually been abandoned. Contemporary ethics of technology is built on this recognition, implying that there are choices possible between better or worse technical alternatives, which makes room for ethical reflection. Likewise, development ethics only became possible when the idea of a unilinear and deterministic development process was abandoned. Thirdly, ever since the empirical turn in philosophy of technology (Kroes and Meijers 2000), the dominant view is that technologies should be studied as concrete phenomena in concrete empirical contexts, with attention for the differences between them. This, of course, requires collaboration with the social sciences, and ethics of technology is thus highly interdisciplinary. Again, this is similar to the situation in development ethics. To conclude: in the past three decades or so development ethics and ethics of technology have each evolved into recognizable disciplines which have – at least to some degree – become institutionalized; For each there are textbooks, courses, conferences, journals and associations. And the direction in which these young disciplines have matured, makes them quite compatible,[6] and has paved the way for more and more systematic ethical reflection on 'technology and development' in the future.

One could argue that there is currently, however, little explicit and recognizable overlap between these two disciplines. Overall, it seems, ethics of technology – and philosophy of technology more broadly – rarely addresses technology in the context of poverty reduction or development in the global South, whereas development ethics rarely discusses technology. There thus seems to exist a gap between both domains of applied ethics, so several sources indicate – although the evidence is fragmented and inconclusive. Selinger (2007), for example, concludes that "philosophers of globalization" hardly discuss technological devices or systems, and even when they do "analysis is restricted to the outcomes or general features of technological practice. These high levels of abstraction allow [only] for general points about responsibility and well-being" (p. 14), and not for an in-depth treatment of the role of technology (let alone engineers) in these issues. And unfortunately "philosophical analysis of emerging development issues involving technology transfer remains scarce" (Selinger 2009, p. 377). The literature on global justice and development ethics, so I also concluded some years ago, at most mentions technology occasionally, in a superficial and simplistic way (Oosterlaken 2009b).[7]

---

[6]It might be that all maturing areas of applied ethics go through a similar evolution – I'm just able to say something about those two.

[7]What is not helping is perhaps that "the tendency within the development literature" in general has been to overlook the active, shaping role of "professional technological agents" like engineers

To my knowledge it is vice versa also the case that the development context is hardly discussed within the literature on ethics of technology. In engineering ethics, more specifically, work explicitly responding to the context of developing countries also seems scarce, although there are of course exceptions (see e.g. Schlossberger 1997; Harris 1998; Luegenbiehl 2010).

An interesting area of technology to zoom in on is ICT, as on the one hand computer and information ethics is one of the major sub-disciplines in ethics of technology, and on the other hand in the past one or two decades ICT has become very popular within development research and practice – to the degree that it is recognizable as a separate field of 'ICT4D', with its own journals and conferences. Recently Dearden (2012) systematically reviewed the academic literature on ICT for development, and development studies more broadly, for their ethical content. He found a "large number of papers [more than 130] that include terms such as ethic, ethics or ethical", but "many papers included the term ethics in the title of a reference but not in the main body of the paper." In a more casual way he also looked at some computer ethics journals like *International Review of Information Ethics* and *Ethics and Information Technology*, but found that these in turn hardly ever discuss ICT4D. Again there are (apparent) exceptions to this lack of attention for ethics within scholarly work on ICT4D. Publications that explicitly address ethical aspects of ICT4D are e.g. articles by communication scholars Hacker and Mason (2003) and ICT4D scholar Unwin (2010). They both start, however, with highlighting – each in their own way – an ethics gap in the literature.[8]

Now this gap is somewhat surprising, as technology has been such an important, integral part of the development endeavor ever since the first technology transfers after WWII took place. One could object that it only *seemingly* exists, as many scholars reflecting on ethical and value aspects of development come from the social sciences or other areas of philosophy – such as well-being research or feminist theory – and do not explicitly associate themselves with 'development ethics' as a specialized philosophical sub-discipline (Gasper 2012). The same holds for ethics of technology. Different approaches, focal points and sub-disciplines exist, and the field is not unified. As with development ethics, scholars working on ethics of technology come from very different disciplinary backgrounds and may not always self-identify as such (Franssen et al. 2009). The term 'ethics' may therefore not figure explicitly in their work. If one takes the time to look for it, one can indeed find work on technology and development that takes a critical stance and implicitly raises an ethical issue or takes an ethical position. For example, Leach and Scoones (2006) distinguish three different ways to conceptualize the relation between technology

---

(Wilson 2008) – which arguably implies that the relevance of engineering ethics is not going to be self-evident to this discipline.

[8]Noteworthy is also an extensive edited volume by Capurro et al. (2007) on African information ethics. However, this seems to take – as Unwin (2010) also notices – a rather permissive stance on what counts as ethics, broadening it to include all empirical work on the social consequences of ICTs.

and development, criticize two of them and defend one. They never explicitly mention ethics, but their discussion fits in very well with those taking place within development ethics on the means and overall ends of development.

Although I consider the pamphlet by Leach and Scoones to be very relevant and interesting as it is, I think in general there is an opportunity to take more advantage of the best that ethics of technology and development ethics have to offer. Making ethical deliberation explicit and drawing on existing ethical literature could increase the depth, scope and quality of such critical reflections. Moreover, arguably this invisibility of ethics in the technology and development literature reflects a general lack of awareness that such practices raise ethical and value issues, which means that some questions that should be asked are currently probably insufficiently put on the table or agenda.

8.4   Filling the Gap: The Capability Approach as a Case

In short, I would thus like to see more work that is explicitly working towards an ethics of technology and human development. However, as said, to make development ethics relevant to engineers, and to make its application to technology more than superficial, it is important to thoroughly explore its connections to engineering ethics, to ethics of technology, and even philosophy of technology more broadly. This can be illustrated by a discussion on the application of the capability approach to technology, an approach that is – as mentioned – very influential within development ethics. By focusing on the capability approach, I do not mean to say that it is a panacea for creating an ethics of technology and development. One omission in the capability approach literature is for example that it has not paid much attention to questions of responsibility (Robeyns 2011), while this is a very important topic within engineering ethics and ethics of technology more broadly. More is thus needed for arriving at a comprehensive ethics of technology and human development.

Yet a development ethics approach that attaches central moral importance to human capabilities offers at least a fruitful angle, as intuitively there seems to be a strong link between technical artifacts and human capabilities.[9] It is therefore not surprisingly that in the past years the capability approach has increasingly been applied to technology and especially to ICT4D (Oosterlaken 2012a). It has been used to criticize the understanding of development that is displayed in 'mainstream' ICT4D practice – arguing for example that it should become less resource-oriented and more respectful of people's agency (see e.g. Zheng 2007; Kleine 2013). The

---

[9]I have clearly not been the only one perceiving this and in recent years philosophers of technology have expressed, clarified and discussed this idea in different ways (i.e. Illies and Meijers forthcoming; Van den Hoven 2012; Lawson 2010). See Sect. 7.5 for a further elaboration on the relation between technology and human capabilities.

capability approach is also increasingly used in empirical work on ICT4D, for example to assess how well ICT4D projects contribute to human development (see e.g. Vaughan 2011; Grunfeld et al. 2011).

However the capability approach, so Robeyns (2005, p. 94) has pointed out, "is not a theory that can explain poverty, inequality or well-being; instead, it rather provides a tool and a framework within which to conceptualize and evaluate these phenomena. Applying the capability approach to issues of policy and social change will therefore often require the addition of explanatory theories." Indeed, as Zheng (2007, p. 10) has noted, "the capability approach offers little about understanding details of technology and their relationship with social processes", or about understanding details of how technology and human capabilities are related. Applying the capability approach without using additional 'technology theories' from fields like science and technology studies (STS) and philosophy of technology thus means that the technology in question can and will then only be discussed in a generalizing or superficial way; it remains a black box. There would, for example, be an important limitation on one's ability to explain the outcome of an ICT4D project. Without such additional 'technology theories' one would not be able to investigate if the choice of the technology, or the way in which it was designed, or its embedding in socio-technical networks, plays an explanatory role in achieving the project outcomes, or can be criticized from an ethical perspective. One might not even think of asking these things, as one might not even be able to fully see their relevance.

If one were, for example, to learn from work in the area of ethics of technology that technical artifacts tend to be value-laden, and that value-sensitive design is a possible way to constructively deal with this, avenues would open up for expanding well-being and agency through 'capability sensitive design' (Oosterlaken 2009a, 2014). Likewise, Zheng and Stahl (2011, 2012) have extensively discussed that combining the capability approach with critical theory, as employed in STS and information systems research, would enable the capability approach to come to grips with the power dimension of technology, which is of course relevant considering the capability approach's commitment to empowering people to lead the lives they have reason to value. Acknowledging the need for technology theories and design approaches to supplement the capability approach of course raises the questions *which ones* could fruitfully do so. This will of course in part depend on one's purpose, but also on things like the general merits of these theories and approaches. There is no one single way to 'operationalize' the capability approach in the domain of technology and design.

Making a connection between the capability approach and some technology theory or design approach may sometimes not only be needed to 'operationalize' the capability approach, but may actually also be beneficial for the technology theory or design approach in question. Zheng and Stahl (2011), for example, assert that the capability approach has added value for critical theory. They feel that critical theorists sometimes get stuck in their attempt to "debunk positive myths" about technology by continuously pointing out how technology is implied in the distribution of power and sometimes even in outright oppression. The capability approach, however, "by seeing ICT as a means to development and asking questions about what conversion factors need to be in place to facilitate the achievement of potential freedom that technology provides" (p. 77) provides an antidote to that tendency. It does so by drawing attention to the potential positive role of technology in the expansion of valuable capabilities. Likewise, the capability approach has arguably something of value to offer to the movements promoting participatory design (Oosterlaken 2009a), inclusive/universal design (Oosterlaken 2012b) and appropriate technology (Oosterlaken et al. 2012). In all these cases, generally speaking the added value of the capability approach lies – not surprisingly – in its ability to facilitate a coherent and systematic normative reflection on the values at stake in the approach in question. Values like agency, justice and well-being become more tangible for engineers and designers by relating them, as the capability approach does, to concrete capabilities and the conversion factors relevant for certain technologies. At the same time, this facilitates a connection between technology and design projects and wider ethical debates about justice and development.

With respect to the choice of technology theories and design approaches supplementing the capability approach, two general issues are worth highlighting. The first is that it may not be inconsequential which 'supplements' one chooses. Although making such a choice will generally speaking be unavoidable to 'operationalize' the capability approach, it may sometimes also be a choice that becomes a topic of disagreement and debate. Robeyns (2008) has shown that one may arrive at different capability analyses or normative evaluations of certain gender cases, depending on whether one supplements the capability approach with a conservative or feminist gender theory. It is thus not only the capability approach that does all the work in such analyses. Something similar holds when applying the approach to technology cases.

For example, when reflecting on technology and the good life, what one sees through the lens of the capability approach will also depend on which of these filters one adds to this lens: a 'pluralist technology theory' or a 'system/network view of technology.' Both acknowledge that technology is not fully neutral towards the good life. Yet the former holds that a myriad of technologies and technological practices exist side-by-side, giving users lots of choices. In combination with the capability approach this seems to lead to the conclusion that in general technology ought to be promoted, as the availability of more technologies will mean that people will have more capabilities, and in that way are empowered to realize their own view of the good life. According to the latter, however, many technologies only function within larger socio-technical systems or networks, which may be more or less restrictive. This perspective is much more pessimistic about the choice that users have to switch between technologies and practices, and in combination with the capability approach would draw attention to the realistic possibility that such systems/network may expand some capabilities while diminishing others (Oosterlaken 2013).

The second general issue is that one should of course be aware that different theories and approaches may, in more or less subtle ways, not be compatible – for example, because underlying assumptions or the understandings of certain notions are clashing. For example, one can draw on Actor-Network Theory (ANT), which is a central theoretical perspective within STS, in order to explain how technology can be given a place in the philosophical ontology underlying the capability approach (Oosterlaken 2011). There are interesting parallels that one could draw between the body of thought of Sen as founder of the capability approach on the one side, and Latour as founder of ANT on the other side (Kullman and Lee 2012). Yet despite these parallels, a salient question is still whether the specific concept of 'agency', which is very central in both ANT and in the capability approach, is actually understood or used in the same way by both thinkers. That is not the case, as autonomous human intentions are very important in the work of Sen, whereas Latour's concept of agency is by and large mechanistic. Another question that one could ask is whether social structures have causal powers above and beyond those of the individuals constituting them. According to some, Sen seems to assume that they do (Smith and Seward 2009), yet ANT denies this. In short, certain insights from ANT are certainly useful if one wishes to give technology a place in the capability approach, but one needs to be aware that one should sometimes explicitly drop or selectively accept certain parts of either view to ensure coherence (Oosterlaken 2013).

8.5 Theorizing the Technology: Human Capabilities Relationship

The previous question argued for the importance of thoroughly thinking through the connections between the capability approach and technology theories. One of the most fundamental questions that arises when applying the capability approach to technology is the question of the nature of the relation between technology and human capabilities. Answering it requires, of course, that one gains a good understanding of both the nature of human capabilities and the nature of technology. One challenge here is that there is no agreement on or single dominant view concerning the nature of technology. Throughout history and in different disciplines 'technology' has been defined and understood in a range of ways, for example as a product, a process or a form of knowledge (Mitcham and Schatzberg 2009). This chapter is not the place to discuss this in any detail, but broadly speaking the view of technology that I would like to propose is that it concerns a set of material artifacts, or systems of such artifacts, designed to perform a certain function. On the nature of human capabilities one could of course also have different views, but within the capability approach the concept has been understood in a quite specific way, namely as having a relational and contextual nature. This means that – in addition to personal characteristics and resources – institutions, additional infrastructure, cultural norms and practices, personal characteristics and so on all matter as inputs, or alternatively as constitutive elements.

Based on this starting point, I would like to argue that both the details of design and the socio-technical embedding of technical artifacts are relevant factors in the expansion of human capabilities. Understanding the relation between technical artifacts and human capabilities requires us thus to regularly move back and forth between 'zooming in' and 'zooming out':

- 'Zooming in' allows us to see the specific features or design details of technical artifacts;
- 'Zooming out' allows us to see how exactly technical artifacts are embedded in broader socio-technical networks and practices.

The latter perspective would also need to take cultural norms, collective usage practices and other 'soft', non-material factors on board. If one were to adopt the view of technological determinism,[10] one would arguably be inclined to adopt mainly or exclusively the zooming-in-perspective. If one were to adopt the view of social determinism, one would arguably be inclined to adopt mainly or exclusively the zooming-out-perspective, be it that one would speak of social instead of socio-technical networks, structures and practices. Just like much contemporary work in philosophy of technology, I thus propose to steer a middle course, one could say, between the extremes of technological determinism and social determinism.

The view just sketched suggests two basic, complementary strategies if for example a development organization aims to effectively expand the human capabilities of marginalized groups with the help of technology. On the one hand one may want to make sure – to the degree possible – that the introduction of a certain technical artifact is accompanied by appropriate changes in the surrounding socio-technical networks (Oosterlaken 2011). On the other hand one may make the design of a technical artifact appropriate – to the degree possible – for the relevant socio-technical network or usage environment as it exists (Oosterlaken 2012). When adopting the capability approach as a value-based perspective that can guide the design of technical artifacts ("capability sensitive design"), this should ideally be based on a thorough understanding of the multiple and complex relations between technical artifacts, human capabilities, and the socio-economical-physical environment in which both humans and artifacts are embedded. Of course both strategies should preferably be combined. Contra Robbins (2007) I thus hold that reflective engineers should hold *both* designs and socio-technical systems in view.

A simple case study of a development project introducing podcasting devices in a rural area in Zimbabwe can illustrate this basic idea. The livelihoods of people in this area mainly depend on subsistence farming. The podcasts recorded and distributed were mainly about cattle management, and meant as a supplementation of traditional agricultural extension services. It concerned a project executed by

---

[10] With technological determinism I do not mean the view that technology develops autonomously, without human influences, but the view that "the physical materiality of technology plays a [determining] causal role in social change" and its social impact (Smith 2006).

Practical Action, a development organization that has its roots in the appropriate technology movement. Hence ample of attention was paid to the technology choice and the details of design. For example, a principled choice was made for podcasting as a voice-based technology, as a substantial number of people in the area are illiterate. In terms of the capability approach: the existing personal conversion factors were for many people such that a text-based technology would not lead to the expansion of human capabilities. A 'design change' that was made later on in the project was that the loudspeakers for the mp3 players were replaced with headphones, as it became clear that this would fit in with existing cultural practices.

At the same time, the project paid quite a lot of attention to fostering new cultural practices that would contribute to the project goals, such as collective listening and discussion sessions, and additional demonstration meetings. However, some relevant aspects of the broader socio-technical system were beyond the control of the development organization. For example, it turned out that these rural farmers could not get access to some medicines recommended by the podcasts, which of course meant that the lessons learnt by the podcasts could not all be fully put into practice. Although the project was never set up with the capability approach in mind, it can be argued that it did have an impact on the human capabilities of the individuals living in the region, and that this impact depended both on the technical artifacts and its design, and on the broader socio-technical networks and practices in which these were embedded (Oosterlaken et al. 2012).

8.6 Conclusion

In this chapter I have argued that in a globalized world a need exists to develop an 'ethics of technology and human development'. In order to give this substance and depth, it should draw on both development ethics and ethics of technology – two relatively young fields of applied ethics which currently have little overlap, but show potential for cross-fertilization. I have illustrated this idea by discussing the application of the capability approach, a prominent conceptual and normative framework within development ethics, to technology. It was shown that insights from fields like science and technology studies (STS) and philosophy of technology are needed in order to make the capability approach relevant to engineering. As such, this chapter makes a small theoretical contribution towards creating an ethics of technology and human development.[11]

---

[11]As said, this still leaves open the question of how to best teach such an ethics, or the capability approach more specifically, to engineers. Experiences with and reflections on teaching the capability approach to engineers can be found in (Boni et al. 2012; Castro-Sitiriche et al. 2012; Frey et al. 2012). See also Frey's chapter titled *Training Responsible Engineers for Global Contexts* in this book.

References

Boni, A., & Perez-Foguet, A. (2008). Introducing development education in technical universities: Successful experiences in Spain. *European Journal of Engineering Education, 33*(3), 343–354.

Boni, A., MacDonald, P., & Peris, J. (2012). Cultivating engineers' humanity: Fostering cosmopolitanism in a technical university. *International Journal of Educational Development, 2012*(32), 179–186.

Brown, A. E., & Grant, G. G. (2010). Highlighting the duality of the ICT and development research agenda. *Information Technology for Development, 16*(2), 96–111.

Capurro, R., Britz, J., Hausmanninger, T., Nakado, M., Weil, F., & Nagenborg, M. (2007). African information ethics in the context of the global information society. *International Review of Information Ethics, 7,* 1–353.

Castro-Sitiriche, M., Papadopoulos, C., Joseph Frey, W., & Huyke, H. (2012). *Sustainable wellbeing education in engineering.* Paper read at 2012 IEEE international symposium on sustainable systems and technology, 16–18 May 2012.

Crocker, D. A. (2008). *Ethics of global development: Agency, capability, and deliberative democracy.* Cambridge: Cambridge University Press.

Dearden, A. (2012). *See no evil? Ethics in an interventionist ICTD.* In Proceedings of the international conference on information and communication technologies and development, IEEE/Institute of Electrical and Electronics Engineers.

Dower, N. (2008). The nature and scope of development ethics. *Journal of Global Ethics, 4*(3), 183–193.

Franssen, M., Gert-Jan, L, & Van de Poel, I. (2009). Philosophy of technology. In E. N. Zalta (Ed.), *Stanford encyclopedia of philosophy* (spring 2009 edition). Stanford: Stanford University.

Frey, W. J., Papadopoulos, C., Castro-Sitiriche, M. J., Zevallos, F., & Echevarria, D. (2012). *On integrating appropriate technology responsive to community capabilities: A case study from Haiti.* In 119th ASEE annual conference and exhibition. San Antonio, Texas.

Gasper, D. (2012). Development ethics – Why? What? How? A formulation of the field. *Journal of Global Ethics, 8*(1), 117–135.

Grunfeld, H., Hak, S., & Pin, T. (2011). Understanding benefits realisation of iREACH from a capability approach perspective. *Ethics and Information Technology, 13*(2), 151–172.

Hacker, K. L., & Mason, S. M. (2003). Ethical gaps in studies of the digital divide. *Ethics and Information Technology, 5,* 99–115.

Harris, C. E. (1998). Engineering responsibilities in lesser-developed nations: The welfare requirement. *Science and Engineering Ethics, 1998*(4), 321–331.

Harris, C. E., Pritchard, M. S., & Rabins, M. J. (2000). *Engineering ethics – Concepts and cases.* Belmont: Wadsworth/Thomson Learning.

Illies, C., & Meijers, A. (2014). Artefacts, agency and action schemes. In P. Kroes & P. P. Verbeek. (Eds.). *Technical artefacts and moral agency.* Dordrecht: Springer.

Kandachar, P., Diehl, J. C., Parmar, V. S., & Shivarama, C. K. (2011). *Designing with emerging markets – design of products and services (2011 edition.* Delft: Delft University of Technology.

Kleine, D. (2013). *Technologies of choice?: ICTs, development, and the capabilities approach.* Cambridge, MA: MIT Press.

Kroes, P., & Meijers, A. (eds.). (2000). The empirical turn in the philosophy of technology. In C. Mitcham (Ed.). *Research in philosophy and technology*, Vol. 20. Amsterdam: JAI/Elsevier

Kullman, K., & Lee, N. M. (2012). Liberation from / liberation within: Examining one laptop per child with Amartya Sen and Bruno Latour. In I. Oosterlaken & J. Van den Hoven (Eds.), *The capability approach, technology & design.* Dordrecht: Springer.

Lawson, C. (2010). Technology and the extension of human capabilities. *Journal for the Theory of Social Behaviour, 40*(2), 207–223.

Leach, M., & Scoones, I. (2006). *The slow race; making technology work for the poor.* London: Demos.

Luegenbiehl, H. C. (2010). Ethical principles for engineers in a global environment. In I. Van de Poel & D. E. Goldberg (Eds.), *Philosophy and engineering – An emerging agenda*. Dordrecht: Springer.

Mitcham, C., & Schatzberg, E. (2009). Defining technology and the engineering sciences. In A. Meijers (Ed.), *Philosophy of technology and engineering sciences*. Amsterdam: Elsevier.

Moskal, B. M., Skokan, C., Munoz, D., & Gosink, J. (2008). Humanitarian engineering: Global impacts and sustainability of a curricular effort. *International Journal of Engineering Education, 24*(1), 162–174.

Nussbaum, M. C. (2000). *Women and human development; the capability approach*. New York: Cambridge University Press.

Oosterlaken, I. (2009a). Design for development; a capability approach. *Design Issues, 25*(4), 91–102.

Oosterlaken, I. (2009b). *Human development and technology: Some observations on the responsibility of engineers and the role of (development) ethicists*. In 8th international conference of the international development ethics association (IDEA), 2–4 Dec 2009. Valencia, Spain.

Oosterlaken, I. (2011). Inserting technology in the relational ontology of Sen's capability approach. *Journal of Human Development and Capabilities, 12*(3), 425–432.

Oosterlaken, I. (2012a). The capability approach and technology: Taking stock and looking ahead. In I. Oosterlaken & J. Van den Hoven (Eds.), *The capability approach, technology and design*. Dordrecht: Springer.

Oosterlaken, I. (2012b). Inappropriate artefact, unjust design? Human diversity as a key concern in the capability approach and inclusive design. In I. Oosterlaken & J. Van den Hoven (Eds.), *The capability approach, technology and design*. Dordrecht: Springer.

Oosterlaken, I. (2013). Taking a capability approach to technology and its design – A philosophical exploration (doctoral dissertation). In P. Brey, P. Kroes & A. Meijers (Eds.). *Simon Stevin Series in the Ethics of Technology*, Vol. 8. Delft: 3TU.Centre for Ethics and Technology.

Oosterlaken, I. (2014). Human capabilities in design for values. In J. Van den Hoven, I. Van de Poel, & P. E. Vermaas (Eds.), *Handbook of ethics, values and technological design*. Dordrecht: Springer.

Oosterlaken, I., Grimshaw, D. J., & Janssen, P. (2012). Marrying the capability approach, appropriate technology and STS: The case of podcasting devices in Zimbabwe. In I. Oosterlaken & J. Van den Hoven (Eds.), *The capability approach, technology and design*. Dordrecht: Springer.

Passino, K. M. (2009). Educating the humanitarian engineer. *Science and Engineering Ethics, 2009*(15), 577–600.

Robbins, P. T. (2007). The reflexive engineer: Perceptions of integrated development. *Journal of International Development, 2007*(19), 99–110.

Robbins, P. T., & Crow, B. (2007). Engineering and development: Interrogating concepts and practices. *Journal of International Development, 2007*(19), 75–82.

Robeyns, I. (2005). The capability approach – A theoretical survey. *Journal of Human Development, 6*(1), 94–114.

Robeyns, I. (2008). Sen's capability approach and feminist concerns. In S. Alkire, F. Comim, & M. Qizilbash (Eds.), *The capability approach: Concepts, measures applications* (pp. 82–104). Cambridge University Press: Cambridge.

Robeyns, I. (2011). The capability approach. In E. N Zalta (Ed.), *Stanford encyclopedia of philosophy*. Stanford: Stanford University.

Schlossberger, E. (1997). The responsibility of engineers, appropriate technology and less developed nations. *Science and Engineering Ethics, 1997*(3), 317–326.

Selinger, E. (2007). Technology transfer: What can philosophers contribute? *Philosophy & Public Policy Quarterly, 27*(1/2), 12–17.

Selinger, E. (2009). Towards a reflexive framework for development: Technology transfer after the empirical turn. *Synthese, 138*(3), 377–403.

Sen, A. (1999). *Development as freedom*. New York: Anchor Books.

Smith, M. L. (2006). Overcoming theory-practice inconsistencies: Critical realism and information systems research. *Information and Organization, 2006*(16), 191–211.

Smith, M. L., & Seward, C. (2009). The relational ontology of Amartya Sen's capability approach: Incorporating social and individual causes. *Journal of Human Development and Capabilities, 10*(2), 213–235.

Unwin, T. (2010). ICTs, citizens and the state: Moral philosophy and development practice. *The Electronic Journal on Information Systems in Developing Countries (EJISDC), 44*(1), 1–16.

Van de Ibo, P., & Royakkers, L. (2011). *Ethics, technology and engineering: An introduction.* Malden/Oxford: Wiley-Blackwell.

Van den Hoven, J. (2012). Human capabilities and technology. In I. Oosterlaken & J. Van den Hoven (Eds.), *The capability approach, technology and design.* Dordrecht: Springer.

Vandersteen, J. D. J., Baillie, C. A., & Hall, K. R. (2009). International humanitarian engineering – Who benefits and who pays? *IEEE Technology and Society Magazine, 28*(4), 32–41.

Vaughan, D. (2011). The importance of capabilities in the sustainability of information and communications technology programs: The case of remote indigenous Australian communities. *Ethics and Information Technology, 13*(2), 131–150.

Whitbeck, C. (1998). *Ethics in engineering practice and research.* Cambridge: Cambridge University Press.

Wilson, G. (2008). Our knowledge ourselves: Engineers (re)thinking technology in development. *Journal of International Development, 2008*(20), 739–750.

Zheng, Y. (2007). *Exploring the value of the capability approach for E-development.* In 9th international conference on social implications of computers in developing countries. Sao Paulo, Brazil.

Zheng, Y., & Stahl, B. C. (2011). Technology, capabilities and critical perspectives: What can critical theory contribute to Sen's capability approach? *Ethics and Information Technology, 13*(2), 69–80.

Zheng, Y., & Stahl, B. C. (2012). Evaluating emerging ICTs: A critical capability approach of technology. In I. Oosterlaken & J. Van den Hoven (Eds.), *The capability approach, technology and design.* Dordrecht: Springer.

Zoomers, A. (2008). Rural livelihoods. In V. Desai & R. B. Potter (Eds.), *The companion to development studies.* London: Hodder Education.

Chapter 9
Ethics, Economics and the Environment

Khalid Mir

Abstract Over the past 30 years there has been a significant attempt to think about the possible relation between the disciplines of ethics and economics. One area in which these debates are particularly striking is climate change. In this paper I argue that engineers need to carefully consider the impact their decisions have on the environment and that the standard ethical approach used by economists in evaluating environmental damage – Utilitarianism- is an insufficiently rich approach towards that end. Instead, I outline a different kind of ethic, one that is centered on care and responsibility, and that may prove to be a more useful way for engineers to understand their role in the modern world.

Keywords Climate change • Utilitarianism • Responsibility

9.1 Introduction

At the outset we might ask the questions: what motivates engineers and why, if at all, should they be concerned about ethics? In this paper I want to suggest that engineers play a crucial role in shaping our world and the world future generations will live in. By 'world' I mean both our social and natural environments. So, I think it is more useful to think of engineers as active agents shaping the world we live in rather than viewing them as mere passive technicians always willing to implement ready-made technological solutions. It follows that engineers should be attentive to both the ultimate aims of technology and to the various means of achieving them. Perhaps they should be thinking of the role of technology and design not in terms of "challenging nature" (Heidegger 1977) but, rather, in terms of maintaining and taking care of it.

Of course, it is readily conceded that in reality the scope for exercising that freedom and judgment, as well as the degree to which any one individual engineer can make an impact, depends crucially on the institutional features of

---

K. Mir (✉)
Lahore University of Management Sciences, Lahore, Pakistan
e-mail: kmir@lums.edu.pk

© Springer International Publishing Switzerland 2015
C. Murphy et al. (eds.), *Engineering Ethics for a Globalized World*, Philosophy
of Engineering and Technology 22, DOI 10.1007/978-3-319-18260-5_9

the organization one works for and on industry standards, norms and incentives. Nevertheless, I do think that the most successful professionals will be those who can both clearly see the different dimensions to problems and who have the ability to think of creative solutions to them. Howard Gardner (2007) may well be correct in asserting for that to occur we will need, amongst other things, the capacity to think along ethical lines. At the very least, it is hoped that thinking seriously about ethics will help engineers give greater consideration to working for organizations whose values match their own.

At the very heart of one of our major concerns, namely the environmental problem, is, I contend, a degree of confusion over exactly what *type* of problem it is. If we could determine the nature of the problem we might make some headway towards a solution. However, even prior to that we have to acknowledge that there is in fact a problem of some sort – and a large one at that too! Given that, the structure of this paper is as follows: In Sect. 9.2 I briefly summarize the magnitude of the problem we're facing. From there, I want to move on and try and address the central question in the paper: what type of problem is it? That is the substance of Sects. 9.3 and 9.4. In the final section I offer some remarks by way of a conclusion.

In Sects. 9.3 and 9.4 I suggest that our standard (economic) approaches to conceptualizing the problem neglect some important features or dimensions of the problem. That is why, it is argued, we need ethics.

It is worth emphasizing a more general point here as well: the way in which we think about something often entails specific policy responses, and rules out others. This is where engineers come in. If we think of the environmental problem as a fundamentally technological one, then we can be inclined to see the solution in those terms as well.

Alternatively, we might think that at the root of the matter there is an economic failure. However, if market economies are self-correcting, as many of its advocates maintain them to be, then the problem will be solved on its own (via prices and incentives and the impetus they give to discovery, innovation, improvements in efficiency, and substitutability). Here, it would seem, the only scope of policy is to strengthen market forces. In addition, though, if economics is at the heart of the environmental problem then surely we should use economic analysis to understand and evaluate the environmental impact of various projects? On both these counts, then, it would appear that there is a limited role for ethics in understanding and/or solving the problem and engineers should not concern themselves with the need for ethical deliberations. Given the title of this paper it will come as no surprise that I believe such a view is mistaken.

A third way of thinking about the environment, then, entails taking on board ethical considerations (Gardiner 2004). To what extent we need an explicit moral *theory* to do so is not something I will discuss here. Nor do I broach the interesting possibility of grounding our concern for, and responsibility to, the environment in religion or a sense of the sacred. My aim here is somewhat more limited and in a sense negative in that I'm only concerned with highlighting some of the aporias in our thinking induced by our reliance on an approach to the environment founded on a particular ethic, namely: Utilitarianism. To paraphrase Sartre (Baier 2008), it is often as important to look at what is missing in an explanation (or theory) as it is to take note of what has been explicitly stated. The point is this: the choice, modification and development of technologies – all things that reflective engineers should be thinking about – embody ethical considerations, not least because they impact the shape of the world we live in, the quality or goodness of our lives, and the distribution of that goodness.

9.2 The Scale of the Problem

It is now widely recognized that one of the most pressing problems we face, a problem that is global in its dimensions, is the degradation of the environment. Its manifestations are manifold: the depletion of the ozone layer, global warming, acid rain, air pollution, soil erosion, the shrinking of Arctic sea ice, a reduction in biodiversity and the contamination of ground water supplies by toxic pollutants. It is also now abundantly clear that accelerated economic activity is the major cause of these changes. However, so far the vast majority of the damage has been the result of economic activity from only a small number of countries – perhaps as few as one fifth of all countries, those that constitute the 'developed' world.

This last point is worth bearing in mind since if the rest of the world follows the same consumption and production patterns as these countries, and if population growth and distribution trends continue, what to us now appears as only a possible disaster will ultimately take the form of an *inevitable disaster*. On a per capita basis developed countries are currently responsible for four times the emissions of developing countries and up to ten to fifty times that of the low-income countries (Royal Society 2012). Nowhere, then, is there a greater need for engineers and policy makers to think more carefully about technology and its impact.

For the purposes of this paper we are only focusing on the global warming aspect of the looming environmental crisis, though in reality there are other crucial issues at stake as well. Though there isn't unanimity with regards the predictions, some of them are quite stark (Broome 1992). Uncertainty is, in fact, a central feature of climate change. In reality we have a cascade of uncertainties since we are not sure how fast emissions will increase under the various projected scenarios or how they will impact temperature change. The change in temperature itself depends on, amongst other things, the level of greenhouse gases (GHGs) in the atmosphere (in particular, carbon dioxide emissions) and those levels, in turn, depend on both how much is released into the atmosphere and how much is absorbed (by forests and the seas). Furthermore, there is some uncertainty over how those temperature changes translate into 'damages'.

If we look at trends in world CO_2 concentration levels we see that they were at 280 ppm (parts per million) in 1769 and 380 ppm now (and increasing by two points every year). It is estimated that 50 % of these emissions are not absorbed and it seems perfectly reasonable to extrapolate from current trends that the absorptive capacity itself will continue to decline with time.

It is worth noting that non-CO_2 emissions contribute up to 45 % of the overall greenhouse effect (Mackay 2009). If we look more generally to GHGs – which include CO_2 but also nitrous oxide and methane amongst others – and factor them in our evaluation, then the current CO_2 level can be thought to be equivalent to 430 ppm. If we stabilize at 450 ppm temperatures are expected to rise, on average, anywhere between 1 °C and 3.8 °C (with 90 % confidence). It is worth emphasizing that these are average figures with the reality for different communities/countries likely to be a lot more varied.

If we take 'moderate' steps to curtail emissions temperatures may still rise by 2–3 °C. Even then, there is the possibility that things will turn out to be much worse. The Intergovernmental Panel on Climate Change (2007) also states that there is a 5 % chance that temperatures will increase by more than 8° and a 1–2 % chance that they will do so by more than 10 °C. These figures highlight the fact that we're not even sure of the magnitude of changes in temperature because our estimates are based on observed data and extrapolation, not on the occurrence of rare events. There may be 'thresholds' which means that the impact is only linear in temperature increases over some range. In that case extrapolation is not justified.

If we stabilize at 550 ppm, however, then temperatures are likely to rise by anywhere in the region of 1.5 °C–5.2 °C (again, with 90 % confidence). The bad news is that if current trends persist we can expect levels of up to 750 ppm by 2100 (with mean temperatures rising by at least 2.2°, and possibly by up to 6°). CO_2 emissions, unlike methane, hang in the air for 50–100 years, so even a drastic reduction now means we'd still see increases on our current levels. The real question, then, is not whether levels will go beyond 430 ppm but by *how much*? (Roemer 2009)

To highlight just how late we might already be the following estimates should be noted: if emissions fall by 70 % (or by the greater amount, 85 %) by 2050 then there's still a 16–43 % chance (or a 9–26 % chance for the greater reductions) that world average temperatures will rise by more than two degrees Celsius with significant impacts on our lives.

The long view, then, indicates that emissions which were relatively stable over the last millennium started to spike from the industrial revolution onwards. Some philosophers believe that this fact warrants something like a precautionary principle given the possibility of a catastrophe and our levels of uncertainty.

9.3 Ways of Approaching

The burning of fossil fuels and, more broadly, our technological choices, are a crucial issue of our time given the scale of their impacts. But what kind of problem is it and which approach should we take in assessing the impacts of projects on the environment?

It is not my aim to give the false impression that there aren't technological, legal, or economic aspects to the environment problem. After all, we can readily admit that poor technology choices are part of the problem and that better, more efficient technologies are part of the solution. For instance, it is clear that a move away from electricity generated by coal to electricity that is generated using low-carbon technologies or fuels can have significant impacts (Nordhaus 2013). Similarly, it is not without merit to mention that the lack of property rights – both at the global and national levels – as well as their enforcement is a pertinent feature of the problem. Recognition of this point can lead us to a discussion of possible remedies that rely on such legal and political arrangements. Furthermore, as has already been mentioned, we should not deny that the environmental crisis is a result of our economic activity. My main aim in the remaining part of this paper is to question whether it follows from the acknowledgement of the latter point that we should rely on economic approaches to evaluating the environmental impacts of projects. This is something we now turn to.

9.3.1 Economics and Ethics

The standard approach to evaluating the impact of projects consists of what we usually call Utilitarianism and derivatives of it (such as cost-benefit analysis (CBA)). In this section I highlight the main features of this approach and in the following look at some of the problems associated with it. Criticism of Utilitarianism is, of course, nothing new but for our purposes we want to tie in those critiques to a better understanding of the nature of its drawbacks when applied to the environmental problem. In particular, we want to ask: does the standard approach miss out on some ethical considerations?

Climate policy is often explicitly guided by the notion that CBA determines the efficient level of pollution, the point at which the marginal costs of abatement are just equal to the marginal "damages" caused by it. Policies such as "cap and trade" or "carbon taxes" then act as an incentive to consumers and producers to get us to the desired, efficient level. But built into that approach are a number of highly contested assumptions – shared by Utilitarianism – that make us question whether climate policy is based on firm foundations. Firstly, it assumes we can put a value to the various kinds of "damages" (namely: to health, life, and the erosion of eco-systems and bio-diversity). Secondly, it assumes we can aggregate these values across individuals for any one generation. This is equivalent to determining the distribution of costs and benefits or, in other words, the distribution of emission permits in a "cap and trade policy", across countries and firms. Thirdly, it assumes we have a sound way of discounting the "damages" inflicted on future generations.

Utilitarianism, it has often been said, is the dominant framework guiding public policy (Wiggins 2006). Although it has undergone some modifications over time (notably in terms of what is meant by utility) the fundamental structure, if not the content, of the doctrine remains intact. Following Amartya Sen we can factor Utilitarianism into three components for analytical convenience. For brevity, I will concentrate only on the aspects that are relevant to our discussion of environmental values.

1. **Consequentialism**: the impact of a project/policy/ technology should be assessed in terms of its consequences (and only its consequences).

   This rules out, I think, paying attention to things we think are 'intrinsically' right as well as to processes that we have reason to value independently of their consequences. Some have argued that such independent factors can indeed be accommodated for by consequentialism if we broaden our view of 'consequences' (Broome 2004). The right action, it would then follow, would be to maximize the independently derived 'good'. Whether such a way of thinking is ultimately too reductive or not is not something I am going to dwell on, but I do think we should be attentive to the possible limitations implied by this view of things. Another way of stating this is: not everything is a means to an end.

2. **Welfarism**: Not all consequences are to count or to be considered. What matters to the utilitarian are, not surprisingly, people's utilities or welfares (sometimes the word well-being is used interchangeably). To use Sen's terminology: utility is the relevant 'evaluative space' (or the 'informational basis' for our theory)

3. **Sum Ranking:** This suggests how we go from an individual perspective to a social one. In other words, how do we aggregate across individuals? Utilitarianism suggests the principle of sum-ranking, namely: whatever actions or choices maximize the sum of utilities is the 'best'.

Utilitarianism has been critiqued along each of these three dimensions (Williams and Sen 1982) and some of those criticisms will throw into relief the relevance of ethical considerations. The point worth emphasizing here is that it is not clear whether it can, as a theory, capture some of the important ethical features of the environmental problem and it is to this criticism of it that we now turn.

9.3.2 Evaluative Space

We have already said, not unreasonably, that we should be concerned about the impact of a project on individuals' utilities (the criticism of this view, it is recalled, is only as to whether this is the *only* thing we should take into consideration). However, even if we give our assent to this view we would still be left wondering what precisely is meant by this term? For example, is it happiness or pleasure; is it a mental state or does it imply something more objective, such as a determinate ethical way of life? Is it, as the economists would have us believe, actually a vacuous concept, a formal and not a substantive concept? If so, utility is a mere numerical representation of someone's underlying preferences (Broome 1999a)

There are several drawbacks with each of the associated points mentioned above but for our purposes I only want to emphasize a few of them.

The first point to note is that the utilitarian is someone who subscribes to the notion that the different arguments in his utility function are, ultimately, commensurate with one another. In simple terms: if it is asked if utility is one thing or many he must answer 'one'. So, we are to assume that we or any individual can aggregate across various 'goods' by comparing them to get to some overall level of utility. As opposed to this view is the notion that some values may be incommensurable and that there are differing views on the appropriate way to value different things, depending on our ideals, the context of our choices, and our conceptions of the self (Beckerman and Pasek 1997). But an additional problem arises here since even if we agree in principle to the idea that environmental goods should be treated as if they were commodities from which we derive utility or value, it is not at all clear whether in practice we can ascertain what that value is.

Markets usually reveal the relative value of commodities through their relative prices since prices indicate how much we would be willing to sacrifice one good for another. How, then, do people value environmental goods if there is no market for them? One way out of this quandary is to directly ask people how much they actually value the environmental good (such as clean air, the aesthetic value of woodlands etc., etc.). Such an approach may not be very helpful, though, for two reasons. Firstly, in actual surveys individuals tend to respond to such questions as if they were being affronted or being accused of being immoral or bad citizens. From those responses one is led to the conclusion that environmental goods are of infinite value or that the preservation of them trumps all other considerations. There is no need to labor the obvious point that such an evaluation, even if it is a true expression of people's convictions, doesn't get us very far in resolving the often very profound dilemmas and trade-offs involved in our choices. Those difficult choices only make sense against the background of *finite* values; otherwise the solution to the problem is a trivial one.

The second feature we should be cognizant of is that since environmental goods are typically public in nature (unlike private goods where someone's benefit from the good comes at the exclusion of another person benefitting from it) there is the possibility of the free-rider problem undermining the willingness-to-pay approach. In short, if there's a tendency to exaggerate the value of environmental goods there is also the possibility of understating their value.

A second noteworthy point follows from our skepticism towards the word utility. If we recall, for modern forms of utilitarianism utility is not a substantive notion (such as happiness or pleasure) but, rather, the satisfaction of one's desires. Economists are generally wary of thinking about the origins of our preferences and how they're shaped – they're just assumed to be 'given' – or, indeed, of thinking of them as being ordered reflectively by reason or ethics. Here one objection in particular stands out to this view: do individuals truly understand or appreciate the value of the environment? If our notion of utility is desire-satisfaction then it could be argued that a policy that leads to the satisfaction of someone's desires – no matter what they are – should count as an enhancement of their well-being; to argue otherwise, or so the economist/utilitarian might contend, is to open the door to paternalism.

At the heart of Utilitarianism, then, is a particular view of the individual and of human rationality. Individuals know what they want, have perfect foresight, have sovereignty over their choices, and try to maximally satisfy their preferences. To take the discussion in the direction of asking whether there are appropriate ways of evaluating things, ways that relate to social relations, obligations, a sense of solidarity (all of which we may call 'the ethical') is to talk in terms that the standard approach is not partial or inclined to. And to bring in the perspective of human needs, as opposed to desires, is to raise both the general question of what type of person we would like to be as well as the more particular one of whether we need a view of rationality that is broader than the one presupposed in Utilitarianism.

We can summarize the aforementioned points in a simple question: if you prefer a lavish but environmentally-damaging lifestyle because you value it more than the environment, then what is wrong with that? To which the ethicist might reply: your actions impose a cost on other people, those currently living as well as future generations. But what if *everyone* values growth and the lifestyle of high-consumption more than the environmental costs of it? It is not obvious that such preferences are necessarily irrational. Or, alternatively, if there is an externality and the benefits of growth outweigh the costs of pollution then how persuasive is a utility-based argument for restraint? There are, of course, other perspectives, such as that of justice, which maintain that we have a moral duty not to inflict harm on anyone irrespective of the overall level of utility or 'goodness' achieved. It might be said, therefore, that unlike Utilitarianism considerations of fairness, rights, responsibility and justice resist the impulse to 'weigh up' across diverse sets of 'goods'.

9.3.3 Summary

In this section we discussed whether Utilitarianism could adequately take account of our values, such as freedom, or respect for the environment. It was argued that because the view relies on desire-satisfaction it may fail to give full weight to our concerns. This followed from the fact that we may not always have informed desires but even if we do, we may simply not think it appropriate to value nature as if it was just another commodity (and that is what is required of Utilitarianism). Furthermore, the theory doesn't prove very helpful if it turns out that we do, in fact, value greater material consumption over the environment. I mean, after all, it could be argued that it has been precisely such relative valuations over the last 200 years that have resulted in us being in the predicament we're currently in today.

9.4 Sum-Ranking

I want to now turn to a different category of problems associated with what we've been rather loosely calling the 'standard economic approach' and its relevance for evaluating the impact of technologies and projects on the environment. The fundamental issue at hand is, quite simply, which policy/technology/project leads to the best overall outcome for society? I'm going to argue the obvious point that any attempt at answering that must first lay out what is meant by 'the best overall'. It is in regard to this question that Utilitarianism as a theory or a guide to policy is found to be lacking, offering only partial insights. Another way of expressing our concerns is to note two reservations. Firstly, under Utilitarianism there is little explicit recognition of the underlying ethical standpoint that it presupposes. This is a significant shortcoming. Secondly, the theory may fail to take into account important ethical considerations. I address both of these concerns separately in the following two sub-sections.

9.4.1   At t_0

How do we get from facts about individual levels of well-being to the overall level of well-being in any given society (i.e. its social well-being or social welfare)? As is the case with other aspects of climate change we're drawn into a discussion of values. Here the more specific question relates to which 'aggregating principle' we should select. The first thing to note is that in any attempt to aggregate across individuals we have to assume that individual levels of well-being *can* be compared. This hasn't always been agreed to, but if we are going to make any progress in our discussion we should put this complication to one side and ignore any claims it has on our attention.

Built into Utilitarianism (point 3 in Sect. 9.3.1 above) was the notion that it relies on sum-ranking. A social decision system based on this principle holds that at any point in time (t_0) the best overall outcome or state of affairs is one in which the aggregate level of well-being is maximized. Although this view is deeply individualistic and though it depends on a particular view of rationality it shouldn't be concluded that it necessarily promotes selfishness. After all, it is certainly within the bounds of this theory for each individual to give weight to the welfare of other people. For example, the theory, as it stands, doesn't rule out the possibility that individuals have altruistic preferences.

One of the criticisms of this approach, however, is that it doesn't give due consideration to the distinctness of persons. To see this, imagine a scenario in which we shifted resources in such a way that a rich person was made a lot better-off and a poor person only slightly worse-off. Under our approach such a move would be approved of since it would improve the aggregate level of well-being-and that is all that is of relevance.

Can such an approach, then, give due importance to our concerns about distribution and fairness – either in any given society or, taking on board the larger perspective, in the world at large? To try and elaborate upon the last point we can turn to the infamous Larry Summers' memo (Hausman and McPherson 2006). If people in rich, advanced countries value clean air, say, more than additional units of money in comparison with those in poor, less-developed countries, and if the costs of marginal increases in pollution (in terms of foregone wages due to ill-health) are relatively lower in the latter as well, then doesn't it make economic sense to shift 'dirty industries' to those poorer countries since overall welfare would be enhanced? In support of this argument it might be pointed out that since initial levels of pollution are likely to be lower in the developing countries the marginal impact of more pollution will be lower there.

This is the kind of policy that economists would say leads to a Pareto improvement in the sense that benefits and costs are so arranged that at least one party (country) is made better-off without any other being made worse-off. The Pareto principle, then, is an aggregating principle that is not fundamentally concerned about the distribution of welfare between people (countries).

The Pareto principle doesn't seem like a very reasonable aggregating principle, though it is one that is central to how economists think. The Utilitarian aggregating principle, however, is equally – if not more – oblivious of distributional concerns. If resources or production are reallocated in such a way that those who are initially at a high level of welfare are made a lot better off at the expense of those who are initially at a low level then Utilitarianism would endorse such a change if the *overall* levels increased. This is just another way of saying that its advocates aren't overly worried about the distinctness of persons.

This serious neglect of distributional concerns can be clearly seen if we ask ourselves: does each citizen have an equal *right* to a pollution-free environment or is that a good she can trade-off for greater income and therefore welfare? Incidentally, cost-benefit analysis, which is supposed to be a practical embodiment of the theoretical (Utilitarian) approach, also fails in this respect since it typically ignores the distribution of costs and benefits amongst different individuals.

The standard economic approach, it has to be said, deals with these complexities in an unsatisfactory way, which is to say that it does so by ignoring them! The economic approach typically assumes that each generation is represented by a single agent. This at once cuts off all discussions over the very real and significant intra-temporal conflicts between individuals in any given country as well as those between countries. The aggregation problem at this level, then, is not something we will pay further attention to. Instead, let us turn to another dimension of the aggregation problem, one that is inextricably related to time.

9.4.2   Future Generations

In this section I want to suggest that there are in fact two distinct problems related to the aggregation of individuals' utilities over time. This two-fold aspect corresponds to the two drawbacks of the Utilitarian approach noted in Sect. 9.4, namely: on the one hand it doesn't explicitly recognize the ethical standpoint it entails and on the other it fails to take account of important ethical considerations. We briefly look at both of these shortcomings.

9.4.2.1 Discounting

To get the discussion started let us at once note that many projects or choices of technology will have impacts not just on those people who are alive now (at t_0) but also on future generations. The aggregation problem, seen in this light, is how to determine social welfare over time. For simplicity, we ignore complications arising from the fact that we do not know with certainty how the projects or technologies we choose impact those future generations. Part of the environmental problem is, no doubt, that we may face radical uncertainty – which is a distinct category from risk, as Frank Knight pointed out many years back. At best, we may only have a probability distribution over those outcomes. So, in reality our problem is how to aggregate over *expected* outcomes. But let us ignore this aspect of the problem in order to focus on another issue. Let us also, for simplicity, assume that each generation can be represented by a single agent, thus relegating the importance of the kind of issues discussed above (Sect. 9.4.1) to the sidelines.

The general problem facing the social planner is to choose a set of feasible consumption bundles in order to inter-temporally maximize social welfare. We ignore political questions of *how* the aggregating principle is decided upon, though in some sense it is crucial since notions of the individual and the social are implicated in any choice of social welfare function.

The point is that the functional form of the social welfare function embodies our ethical viewpoint, even though we often fail to recognize this. One way of talking about ethics, then, would be to ask what we think are reasonable properties of our aggregating principle (in addition to the processes involved in getting to one).

The Utilitarian view proposes a simple maximization of utilities over time. As it stands this approach has the apparently desirable property of not discriminating between generations, giving each equal weight in its overall assessment of inter-temporal welfare. We may still have some reservations, similar to the ones noted above, namely: it doesn't pay much attention to the distribution of welfare between generations. To see this, think of two distributions. In the first the current generation has very high levels of welfare and the future ones very low levels; in the second distribution assume that both generations have roughly the same levels of welfare. If the total level of welfare under both distributions is the same Utilitarianism doesn't help us distinguish them though from an ethical point of view we might justly be worried about the inequalities expressed in the first distribution.

However, surprisingly economists do not use this form of the social welfare function but, instead, one that discounts future utility. The point to note here is that the discounted function is not symmetric in that it gives greater weight to an increase in a given amount of utility to current generations than it does to a similar increase to future ones. Well might we ask the reasons or justifications for using this particular form? I'm not convinced that we do have reasons or, rather, good reasons for doing so but the main point worth re-iterating is that the discounted form embodies an ethical standpoint that is not explicitly acknowledged. Namely, it claims that the welfare of future generations is of less value than an equal amount of welfare accruing to someone living now.

The choice of the discount rate, then, cannot be made simply on technical grounds. It is perfectly understandable that *individuals* would want to discount future welfare, just as it is that society should discount future *commodities*, but the idea that society should discount future welfare is highly problematic (Beckerman and Hepburn 2007). In fact, writing back in 1928 Frank Ramsey, the distinguished economist, wrote: "Discounting is a practice which is ethically indefensible and arises merely from weakness of the imagination" (Wiggins 2011). Furthermore, in his influential review of climate change Stern (2007) explicitly states that ethics and value judgments play a role in evaluating not only the aggregation of utilities across individuals at any one given point in time, but also in the aggregation of them across generations.

Stern tends to favor a discount rate which is close to 1, which means that in effect people in the future are not discriminated against in our policy analysis simply because they are born much later. However, having said that, it should be noted that questions surrounding the discount rate continue to be a point of fierce contention between economists and philosophers (Broome 1999b).

Before moving on, it is worth emphasizing another point. We almost instinctively assumed that we should maximize inter-temporal social welfare. We are now cognizant of the fact that how we aggregate or weigh up lives is itself a matter of choice, reflection, and ethical judgment. We could, quite conceivably, argue that instead of unreservedly trying to maximize overall welfare we should try and maximize it subject to the constraint that no single generation slips below some adequately defined cut-off level of welfare. Such an ethic, a maxmin strategy, can justifiably be called a "sustainable" approach to the environment problem (Sen and Anand 2000).

9.4.2.2 Missing Persons

In this final section before the conclusion I want to draw attention to another dimension of the aggregation problem, one that is related to time but, I believe, distinct from the one mentioned in the analysis in the preceding section.

Life is full of blank spaces. What I mean by that is not that there was a time before each of us existed and a time after which we will have ceased to exist. I'm not sure to what extent either of those features do or should worry us. Nor do I mean to highlight that stretches of our own lives can sometimes appear to be blank to us. What I mean by that statement is something altogether more mysterious: as a result of our choices of technology or projects some people will exist who otherwise wouldn't have existed and, conversely, some people who might have existed will not.

The question that John Broome (2005) raises, then, is this: should we take into account the well-being of such people in our evaluation of policies/technologies? There may be technical reasons that incline us to think that we shouldn't, namely, the sheer degree of complexity added to our standard approach in doing so. But even if we can second-guess in this regard, should we, *ought* we to include the well-being of such people in our analysis? Again, for simplicity, let us abstract from the possibility that their existence and well-being affects *our* well-being. The question we are asking is of a different order: should the existence of such people and *their* well-being be given weight in our assessment?

I hope the distinction between this point and that made in the previous section is clear. There our concern was over the relative weights assigned to members in a *given* series. The point being raised here, on the other hand, is that there may not be a given series. Broome argues, plausibly, that we ought to be concerned about extreme levels of well-being but beyond that we should be neutral to a whole range of individual well-being of such future people.

However, if the neutrality assumption doesn't hold we are left with an unpalatable conclusion. Global warming will result in the deaths and suffering of many people and we all agree that that is an unequivocally bad thing. But if we take steps to cut back on emissions we will have to sacrifice resources now. The additional point made by Broome is that the world's population will change by following some precautionary steps. Should we take into account the value of population, the well-being of these future people? If we are not neutral and we do in fact do so, it is quite conceivable that this (negative) value outweighs the disvalue of all the killing and that subsequently global warming is a good thing!

9.5 Conclusion: Take Care!

By way of concluding this paper I do not want to reiterate the main points but, instead, draw upon and extend the criticisms made in the last section (9.4.2) as a way of suggesting why Utilitarianism as an approach is ultimately flawed.

At the beginning of this paper I asked the question: why should engineers be concerned about ethics and what type of ethical framework might help them? To the first part of that question we claimed that we (and in particular engineers) stand in a unique relation to future generations and to the environment. Never before have our choices had such a great potential of impacting the future. In fact, given the looming environmental crisis that we have created by of our choices of technology and lifestyles we must face up to the very real possibility that our choices now make future existence itself a precarious condition. If so much is true, we need a radically new form of ethical thinking, since the older versions never countenanced this as a possibility (Jonas 1984).

Our choices today will have an impact on not merely the quality of future people's lives and the environment they inherit, but also on the likelihood of the very existence of future generations. In addition, we should also accept that our greater knowledge of the consequences of our choices places a stronger burden of responsibility on our decisions.

In our discussion above it was argued that Utilitarianism may not be the best way of evaluating and guiding our decisions. Why should we, the current generation, sacrifice our well-being for people we do not even know or who may not even exist? In fact, as was noted (in Sect. 9.4.2) it is not clear that a perspective relying on a utilitarian framework dictates that we necessarily should.

Are there other ways, then, of thinking of this problem? Can we think, perhaps, of future generations having rights (to clean air, say) and does that translate into a set of corresponding obligations on our behalf? Does our ultimate responsibility to them stem from the fact that our choices/policies alter the size and nature of the future population? "We owe it to them," we might say, even though we don't know who 'them' is, how many of them there will be or, indeed, what their desires and goals will be. Can we meaningfully talk about people's rights when they're not even around to make claims on us and what type of political arrangements might we conceive to ensure that they have some form of current representation? You can see here, already, that I think we need to shift our discussion of the problem away from the standard approach, with its heavy historical debt to economics, to one that encompasses political analysis.

I think the way out of our predicament, if there is a way out, depends on more than a reinvigoration of political commitment though. In the final analysis it will require of us to think about ethics, values and their supporting culture more seriously (Keat 1994). What eventually grounds our obligations to future generations? This is not the same as asking the difficult question of the *extent* of our obligations, which may depend on the size of the future population but, rather, to ask whether we have a fundamental obligation to, at the very minimum, ensure that there will be people around like us, with the same responsibilities to *their* future generations. Future generations it might be said have, like us, the right to form their goals and preferences. However, those future people cannot do that if they do not exist in the first place or if there is nothing left of the environment for *them* to pass on.

One way out of the impasse, then, would be to imagine ourselves as part of a continuous moral community (Baier 2010). Here we are, part of the human chain, between past and future. We stand in an asymmetric relation to the future generations, one in which they are dependent on us. But we, too, were dependent on others before us. And perhaps it is this system of inter-relations, of mutual dependencies, that offers the way forward. Instead of thinking of ourselves as autonomous, self-interested utility maximizers – which is the foundations of the Utilitarian view we have been discussing – would it not make more sense to of ourselves as "rational dependent animals" (Macintyre 1999) and to think of our relation to the environment and future generations in terms of stewardship (Arrow 2004): our obligation to preserve, regenerate and renew that which we didn't generate ourselves, but simply inherited? To do so would mean radically rethinking what we mean by words like 'resources.' (Berry 2010) It would also mean an acute awareness that we're nearly out of time.

References

Arrow, K. J. (2004). *Speaking for children and for the future*. Paper presented at the tenth plenary session of the pontifical academy of social sciences, Vatican City, 29 Apr – 3 May 2004.
Baier, A. (2008). *Death and character: further reflections on Hume*. Cambridge, MA: Harvard University Press.

Baier, A. (2010). *Reflections on how we live*. Oxford: Oxford University Press.

Beckerman, W., & Hepburn, C. (2007). Ethics of the discount rate in the stern review on the economics of climate change. *World Economics, 8*(1), 187–210.

Beckerman, W. & Pasek, J. (1997). *Plural values and environmental valuation*. CSERGE Working paper GEC 96–11.

Berry, W. (2010). *What matters? Economics for a renewed commonwealth*. Berkeley: Counterpoint Press.

Broome, J. (1992). *Counting the cost of global warming*. Cambridge: White Horse Press.

Broome, J. (1999a). *Ethics out of economics*. Cambridge: Cambridge University Press.

Broome, J. (1999b). Discounting the future. In *Ethics out of economics*. Cambridge: Cambridge University Press.

Broome, J. (2004). *Weighing lives*. Oxford: Oxford University Press.

Broome, J. (2005). Should we value population? *Journal of Political Philosophy, 13*(4), 399–413.

Gardiner, S. M. (2004). Ethics and global climate change. *Ethics, 114*, 555–600.

Gardner, H. (2007). *Five minds for the future*. Boston: Harvard Business Review Press.

Hausman, D., & McPherson, M. (2006). *Economic analysis. Moral philosophy and public policy*. Cambridge, England: Cambridge University Press.

Heidegger, M. (1977). The question of technology. In K. Farell (Ed.), *Basic writings*. New York: Harper & Row.

Intergovernmental Panel on Climate Change. (2007). Cambridge, UK: Cambridge University Press.

Jonas, H. (1984). *The imperative of responsibility: In search of an Ethics for a technological age*. Chicago: The University of Chicago Press.

Keat, R. (1994). Citizens, consumers and the environment: Reflections on the economy of the earth. *Environmental Values, 3*(4), 333–349.

Macintyre, A. (1999). *Dependent rational animals: Why human beings need the virtues*. Chicago: Open Court.

Mackay, D. (2009). *Sustainable energy without the hot air*. Cambridge, UK: UIT Cambridge Ltd.

Nordhaus, W. (2013). *The climate casino: Risk, uncertainty, and economics for a warming world*. New Haven & London: Yale University Press.

Roemer, J. (2009). *The Ethics of distribution in a warming planet*. Cowles foundation discussion paper, no. 1693.

Royal Society. (2012). *People and the planet (the royal society science policy report 01 (12)*. London: The Royal Society.

Sen, A., & Anand, S. (2000). Human development and economic sustainability. *World Development, 28*(12), 2029–2049.

Stern, N. (2007). *The economics of climate change: The Stern review*. Cambridge, UK: Cambridge University Press.

Wiggins, D. (2006). *Ethics: Twelve lectures on the philosophy of morality*. Cambridge, MA: Harvard University Press.

Wiggins, D. (2011). A reasonable frugality. *The Royal Institute for Philosophy*, supplement 69, 175–200.

Williams, B., & Sen, A. (Eds.). (1982). *Utilitarianism and beyond*. Cambridge: Cambridge University Press.

Chapter 10
Ethics for Construction Engineers and Managers in a Globalized Market

George C. Wang and John S. Buckeridge

Abstract Ethical decision-making is central to the practice of construction engineering and management. This is no more evident than in the twenty-first century, when the construction industry must function in very diverse organizational contexts. Whilst construction companies pursue projects in international markets, many investors are buying or forming joint ventures with domestic companies. New and varied professional attitudes have recently arrived in western markets such as the United States and Australia because construction companies are increasingly employing managers from developing nations to undertake commercial and infrastructure engineering projects. In many developing countries the construction industry is vulnerable to unethical behavior or corruption – vulnerability in part because of differences in culture and managerial systems across countries; and this diversity is manifest in the different perspectives of professional ethics and professional practice. Importantly, professionals in construction engineering must be aware of these differences; however current ethics education for engineering professionals generally lacks global components. In this chapter, emphasis is placed upon professional registration, including mandatory awareness of professional ethics, as an imperative for the welfare of world citizens; discussion on the nature of the construction industry and globalized trends emphasizes why ethics and professional education must be integrated within civil and construction engineering and management curricula. Only then we can anticipate an appropriate educational foundation for professional registration of the international engineer.

Keywords Construction • Globalization • Ethics • Professionalism • Professional registration

G.C. Wang (✉)
Department of Construction Management, College of Engineering & Technology, East Carolina University, Greenville, NC 27858, USA
e-mail: wangg@ecu.edu

J.S. Buckeridge
School of Civil, Environmental & Chemical Engineering, RMIT University, Melbourne, VIC 3001, Australia
e-mail: john.buckeridge@rmit.edu.au

© Springer International Publishing Switzerland 2015
C. Murphy et al. (eds.), *Engineering Ethics for a Globalized World*, Philosophy of Engineering and Technology 22, DOI 10.1007/978-3-319-18260-5_10

10.1 Overview of the Construction Industry

The construction industry is one of the largest industries in the United States (US), on a par with education and health. Along with associated investment and service industries, it contributes about 9 % of the US Gross Domestic Product (GDP) and employs more than 10 million workers. Construction engineering spans design, new construction, rehabilitation, renovation, maintenance of constructed facilities, manufacture and supply of building and construction materials and equipment through to demolition. From 2010 to 2020 an annual growth rate of 2.9 % in construction related jobs in the US is projected, making this the largest increase in employment among all industries. Estimations are that the construction industry will experience one of the largest increases in real output over this period, rising by \$368.7 billion, to reach almost \$1.2 trillion per annum. Investment in the non-residential market is expected to grow 3.2 % per year, with residential investment projected to grow at 7.0 % per year (Henderson 2012; Nunnally 2007).

From a technical point of view, construction is divided into building (vertical) construction and infrastructure (horizontal or heavy) construction. Building construction focuses on homes, schools, hospitals, sky scrapers and shopping centers; while infrastructure construction industry builds highways, bridges, airports, harbors, dams, pipelines, water treatment facilities, power plants etc. From a business and construction market side, the major types of construction projects can be further sub-divided into commercial, industrial, infrastructure, institutional and residential. Those who are in the construction industry may undertake projects for both private and public interests. The industry also interacts with multiple facets of government at the federal, state and local level.

Constructed facilities are becoming increasingly complex from both technical and managerial perspectives. A single construction project may involve a wide diversity of professionals and skilled workers, including engineers, architects, general contractors, multitudes of subcontractors who tend to specialize, numerous manufactures and suppliers. For a construction project, it is normal to have several thousand, or tens of thousands of activities which need to be carefully planned and coordinated and this involves managing quality, costs, schedules, safety, resources, and more importantly, personnel and reputations.

In most instances, the construction process may go through numerous stages with many parties involved. A new infrastructure construction project may have to pass various stages before construction: a project concept approval, feasibility study, design and zoning approval, land purchasing, bid and contract documents preparation, project procurement, building permits application, invitation of bidders, attending pre-bid meetings, bid analysis, review of qualifications, insurance and contract awards.

During construction, the contractor or the owner's contract administrator will manage the project. Contractors, general and/or subcontractors, and the owner, government or private, or independent third party who works for the owner, will conduct quality control and quality assurance respectively. The construction or project manager will call and preside over pre-construction meetings, weekly, intermediate and final inspection meetings, propose or review change orders, prepare a quantity list and payment certificate, manage cost, schedule, safety, quality, resources including materials, equipment, and personnel. Post construction, the owner or contractor may be involved in maintenance and operation, responsible for warranty repairs and rectification.

The complexity of the construction industry has made it a high risk, volatile business, such that only those companies of proven competence, reliability and demonstrated integrity are likely to survive over time. Construction projects are difficult to manage because each project is unique by nature, involving many skills that are non-repetitive and do not lend themselves to assembly type production – and they are dependent upon environmental and natural conditions which are often beyond the contractor's control.

In the US there are currently about 730,000 construction companies amongst some 27 million businesses, where 90 % have less than 20 employees (USCB 2012). Over 80 % of all construction companies in the US are small firms that gross less than $500,000 annually. For every 1,000 firms in operation, 110–130 enter the field each year, with a similar number leaving – the highest rates of entry and leaving of all industries in the US.

Whether large or small, specialized or general, success depends on the ability to manage personnel, cash flow, and safety; control costs, finance work, estimate, and schedule; and maintain an effective quality control system. For the contractor, technical competence alone does not ensure profit, as competition is intense. For a commercial project, a typical gross profit is approximately 5 %; after home office overhead the profit is reduced to 2–3 %.

10.2 Impact of Globalization on the Construction Industry

The world is rapidly globalizing. Globalization refers to the increasingly international relationships of culture, people and economic activity. Most often, it refers to economics: the global distribution of the production of goods and services through the reduction of barriers to international trade such as tariffs, export fees, and import quotas. Globalization contributes to economic growth in developed and developing countries through increased specialization and the principle of comparative advantage. It can also refer to the transnational circulation of ideas, languages and popular culture.

Not surprisingly, rapid globalization has impacted the construction industry as engineers seek work or pursue projects in international markets (IHS 2013). Even in some traditionally domestic areas, engineers and managers have to consider global factors, such as environmental change that demonstrably impacts regions rather than individual countries (Bird 2003). In the construction industry, foreign engineers, designers, and project managers may come to the US to build the nation's infrastructure facilities; meanwhile, foreign investors are buying or forming joint ventures with domestic companies (All Business 2006). Engineers and project managers are increasingly selecting materials or equipment designed and manufactured by foreign countries for construction projects and this will affect every aspect of public welfare and safety. A high profile example is that of the drywall imported to the US from a foreign country that caused thousands of homeowners to become ill (Sawyer 2009).

Ten to twenty years ago, international construction projects were conducted in developing countries by companies from developed countries (Raftery et al. 1998). However, in more recent years, there have been dramatic changes in the international construction market. Construction companies, foreign engineers and project managers from developing countries are entering developed countries in significant numbers. It is not uncommon for the US contractors to find themselves bidding against foreign contractors on domestic projects; e.g. Chinese construction companies started the Alexander-Hamilton bridge renovation project, the largest single-phase project for the New York Department of Transportation (EU 2009), and are building the new Bay Bridge span in California (PRI 2011). Construction companies are also forming joint ventures like the American/Nigerian gas-to-liquids plant contract in the Niger Delta. Thus even if they work in their own country, engineers, managers and professionals in the US need international knowledge, experience, or awareness of a professional code and practice of conduct and ethics to work with international partners. Globalization's benefits have been stated, but there are costs, including how agencies and organizations monitor and assess the professional licensure and registration for foreign engineers and managers who work for US companies in the US or overseas.

In this rapidly changing, globalized construction market there are increasing opportunities for unethical and corrupt management – where the end justifies a means that would otherwise have been inappropriate. These are discussed in forthcoming sections.

In most engineering projects, ethical decision-making involves issues ranging from the selection of a variety of alternatives and uncertain circumstances (persons, materials, methods) to the consideration of consequences that may affect other stages, projects, companies or persons and have economic, legal and social implications beyond the client.

The main type of ethical problems for organizations includes human resource management, equitable and just treatment of employees, conflicts of interest, and division of the loyalty of employees. Customer confidence falls when a company shows lack of respect for its clients, downplays public safety and makes inappropriate use of corporate resources for personal gain (Barringer and Ireland 2006).

10.3 Professional Issues in the International Construction Industry

The construction industry is oft perceived as being full of good, down-to-earth, honest people (Smith 2010). Construction professionals and workers have designed and built remarkable infrastructure, commercial and residential structures globally and have laid the foundations of modern society. However, a globalized and competitive construction market has had some less than desirable side effects.

Every year, the Consumer Federation of America (CFA) and the North American Consumer Protection Investigators (NACPI) collects and analyzes consumer complaints gathered from local agencies, to better understand macro-level consumer trends. In July 2012, feedback from thirty agencies from across the United States ranked the construction industry as "number three" for complaints of "shoddy work, failure to start or complete the job" (CFA 2012).

In general, the construction industry has a reputation for delays, industrial disputes and late payment, threatening the finances of contractors, subcontractors, funders and clients. One of the most expensive cost overruns was that of the Sydney Opera House, which although budgeted at $7 million, cost more than $100 million and took more than a decade to complete, ending with a cost blowout of 1,400 % (Chua 2013). Yet the reputation of any credible player in the construction industry is a key determinant of the quality of work they win, and the terms on which they can do business. Construction company reputation management is thus a critical business issue.

The rapidity of change in the construction industry has made it one of the most vulnerable areas for unethical, unprofessional and corrupt behavior. But it is also vulnerable because investment in construction is often very large with many participants and complicated processes and regulations.

And it is not just in the US where these issues arise: studies of construction practitioners in Australia, South Africa and the UK also highlight significant and rising amounts of corruption (Carroll and Buchholtz 2006). Construction professionals in South Africa, especially contractors, have reputations for unethical behaviour such as collusion, bribery, negligence, fraud, dishonesty and unfair tendering practices over-claiming withholding payment for service delivered (Pearl et al. 2005).

The Global Corruption Report 2008 (TI 2008) states that the World Bank estimates that 20–40 % of investments in the water sector were lost as a result of corruption.

In 2008 Halliburton agreed to pay fines of $559 million to the US government to settle charges on bribes allegedly paid by its subsidiary, Kellogg, Brown and Root, for the construction of a liquefied natural gas plant in Nigeria, from 1996 to mid-2000s (Iriekpen 2009). It is the largest amount that was paid by a US firm in a bribery investigation.

In 2008 Siemens AG of Germany agreed to pay $800 million in fines to the US government to settle investigations involving alleged payments to government officials around the world to win infrastructure contracts. As part of the settlement, Siemens did not admit to bribery but admitted to having had inadequate controls, and keeping improper accounts (SEC 2008).

In 2009 the UK Office of Fair Trading (OFT) announced its final decision to fine 103 construction firms £129 million for engaging in bid rigging, largely in the form of cover pricing, on 199 tenders between 2000 and 2006 (OFT 2010). In 2008 following a 4-year investigation, OFT issued a Statement of Objections against 112 construction firms for alleged bid rigging.

In 2013, SNC-Lavalin Group Inc., a Canadian construction and engineering company agreed to a settlement with the World Bank that excludes it from bidding on bank-sponsored projects for up to 10 years because of its involvement in a Bangladesh bribery scandal (Haggett and Orr 2013).

These examples all happened in international projects. Not surprisingly, expectations in ethical and professional practice differ in different countries due to culture, overall managerial system and other social factors. Indeed, in some regions of the world systematic corruption happens regularly in the construction industry and as such, engineering education must prepare practitioners for how to deal with this and maintain an ethical behavior.

10.4 Differences in Ethics Education and Practices

Globalization's benefits for the construction industry have been stated. However, the cross impact on professionalism and ethics has seldom been addressed appropriately. This includes the impact, (often negative) of the inevitable influence of professional practice on international construction market. To stop all corruption practices, professional ethics education, with international related components, is seen as an essential component of the engineering curriculum (Buckeridge 1994).

10.4.1 Ethics Basics

To be competent in a global economy, the work of an engineer must demonstrate that it is sustainable – i.e. it must include an assessment of the long-term viability of the proposal, taking into account social, environmental, economic and technical dimensions. In the past, this has been addressed through the implementation of triple-bottom-line accounting.

The concept of triple-bottom-line assessment was created by John Elkington in the late 1990s as a means for organizations to demonstrate that they had strategies in place to ensure sustainable growth (Elkington 1997). The three dimensions considered in Elkington's assessment are the environmental context (impact of the proposal on the biological systems operating within a region), the social and cultural context (impact of the proposal on the lifestyle of people who will be affected by the proposal), and the economic context (financial implications of the proposal). It is nonetheless evident that we are gradually using up our planet – particularly the accessible and non-renewable resources (Buckeridge 2008). Any long-term plans for increased development (i.e. increased extraction and conversion or use) of these by any one sector or nation requires a commensurate reduction in use by another.

However, project sustainability that is assessed through triple-bottom-line accounting is deficient in one key dimension – the role of technology.

If the **engineering** is poor, irrespective of how much effort is placed to ensure that the other three parameters are addressed, the proposal will fail. Through "engineering" we address the technical aspects of a proposal – which in turn are a reflection of the design and the materials used. Further, if the structure has a designed life span, provision should be made to consider what should happen to the materials and the site on demolition. In light of this, Buckeridge (1994) introduced the concept of the "4 Es" wherein those charged with ensuring the sustainability of projects must demonstrate that they effectively meet environmental, ethical, economic and **engineering** criteria. "Environmental" comprises all aspects of nature – conservation, and enhancement of ecosystems and geosystems; "ethical" comprises the human dimensions – cultural, social and spiritual; "economic" comprises the financial support including that to cover contingencies such as hazards during (and post) construction; "engineering' is all about good design and ranges from structural components through to selection of suitable materials. Poor design is clearly unsustainable, from the ethical, environmental, economic and engineering perspectives (Buckeridge 2011).

An assessment of whether an act is ethical is a determination of whether the act is good or bad, or right or wrong. In adopting ethical protocols, engineers accept approved guidelines about behavior, i.e. activities that they should seek to undertake or to avoid.

Thus ethics provides specific rules for determining right and wrong in a given situation. In some ways, the law represents a particular society's "codified ethics". However, ethics is generally held as going beyond the law, and in light of this, professional bodies have found it necessary to develop their own discipline specific protocols, or "codes of ethics". It follows then, that when professionals such as engineers develop their code of practice, as well as guidelines for technology and business practice, they must include an ethical framework, or a Code of Ethics (Buckeridge 2011).

Ethics is intimately linked to professionalism and in many respects "professionalism" can be equated to "professional ethics". The Royal Institution of Chartered Surveyors (RICS) Working Committee (2000) states professional ethics involves "giving of one's best to ensure that clients' interests are properly cared for, but in doing so the wider public interest is also recognized and respected".

Ethical problems do not have such easily defined solutions – rather ethical conundrums are usually surrounded by ambiguities, complexity, and ill-defined boundaries, and involve decisions characterized by choice as alternative courses of action are available – hence ethical choices and ethical dilemmas.

10.4.2 Business Ethics

Carroll and Buchholtz (2006) state the characteristics of moral and immoral managers. Moral managers:

- Conform to high levels of ethical or right behavior;
- Conform to high levels of personal and professional standards;

- Exercise ethical leadership – search out where people may be hurt;
- Aim to succeed but only within confines of sound ethical precepts;
- Display high integrity in thinking, speaking and doing; embrace the letter and spirit of the law;
- Consider law as the minimum ethical level, and operate above legal mandates; possess an acute moral sense and moral maturity.

Not surprisingly, these virtues are the key components of any code of ethics (see IEAust 2000). Intentionally amoral managers do not think ethics and business should mix; they consider business and ethics as existing in separate spheres, and think that ethics is not worthy of the "good manager". Perhaps unintentionally, amoral managers ignore the ethical dimension of decision-making – thinking ethically is not part of their job description. These practitioners lack ethical perception or awareness and are unable to sense or care about ethical dimensions. At best they could be construed as well-intentioned but morally casual, or inept.

This is not to say that businesses should not focus on earning profits: they must take advantage of opportunities. But this should not be at the expense of being a good corporate citizen.

As mentioned, many corruption cases are related to ethical issues. Shenkar and Luo (2004) see corruption as an exchange between two parties who influence resource allocation – immediately or in future, involving the use or abuse of public or collective responsibility for private ends. The World Bank Sanctions Committee (2003) defines corruption as "the offering, giving, receiving or soliciting of anything of value to influence the action of an official in the procurement or selection process or in contract execution". The UNDP (2008) defined corruption more broadly as "misuse of entrusted power for private gain". The International Federation of Consulting Engineers (FIDIC) defines corruption as "the misuse of public power for private profit". Behind all corruption there is a malfunctioning managerial system.

10.4.3 Systematic Corrupt Practices and the Root Causes

In the unitary managerial system countries that may be found in very rapidly developing economies, economic development is often commensurate with construction quality failure (Khan et al. 2008). The pace of development, impressive as it may sound, is all too often marred by poor construction quality, attributed to technical issues including construction materials, equipment and techniques, quality control measures. However, the quality issues and failures in these regions have generally not arisen due to pure technical issues – rather the managerial system is the root cause, placing every citizen at risk of becoming a victim of a disaster caused by the failure to adhere to quality standards.

Construction is a large and complicated process in which quality is especially important. Factors affecting quality directly relate to hardware, i.e. construction materials, equipment and manpower. However, an irrational managerial system and various defects derived from the system can influence a person's ethical and professional decisions, which negatively affect construction quality.

The typical features of a unitary managerial system include: (i) functions of all levels of government overlap; lower level government is responsible for higher level government, not for the people; (ii) power is over-centralized at higher levels of government but those in power take on few responsibilities; (iii) lack of effective supervision mechanism. These features are the source of numerous problems: the government decree is obstructed, poor policy decisions are made, extensive corruption, etc. These defects affect each other and contribute to numerous managerial, legal, moral, and ethical ills which ultimately impact negatively upon construction quality. This unfortunately is not just confined to the construction industry but permeates to other sectors of industries as well. Figure 10.1 shows the typical features of the system and its negative impacts on construction quality (Wang 2006). The dotted arrows represent the negative effects that cross-affect construction quality under the unitary managerial system.

An irrational managerial mechanism will result in confusion of governance operation procedures, which can lead to tremendous abuses.

The power of the unitary government is its over-centralization and its obligations are, as a result ambiguously defined at best. This leads to poor decision-making, the confounding of functions between governments and enterprises, collusion and conflict of interests, unethical behaviors, corruption and other social ills.

The flaws inherent in any managerial system influence each other; they produce various ills, which radiate to the entire industry and personnel, damaging the quality of infrastructure construction. It is through the influence of this on people that systemic flaws permeate construction quality. These flaws and their derivatives restrain and severely compromise the technical initiative of any high quality intention; they also degrade effort and skilful execution, as well as the ethical behavior of all involved in infrastructure construction.

10.4.4 Code of Ethics and Compliance

A major aim of a standard code of ethics is "to respect the inherent dignity of the individual" (IEAust 2000).

To regulate professional work ethics, construction companies and professional organizations adopt codes of ethics and compliance policies. In order to prevent inappropriate practice, construction professionals must ensure their goals are compatible with ethical considerations and that their route to meeting these goals is ethically acceptable. They must consider the consequences of their activities for others, including all the stakeholders, before they undertake them.

Politicians are starting to respond, as most appear to want to be associated with fighting corruption. Governments seek to improve the ranking of their countries on international benchmarks, such as the Corruption Perception Index (CPI), for instance.

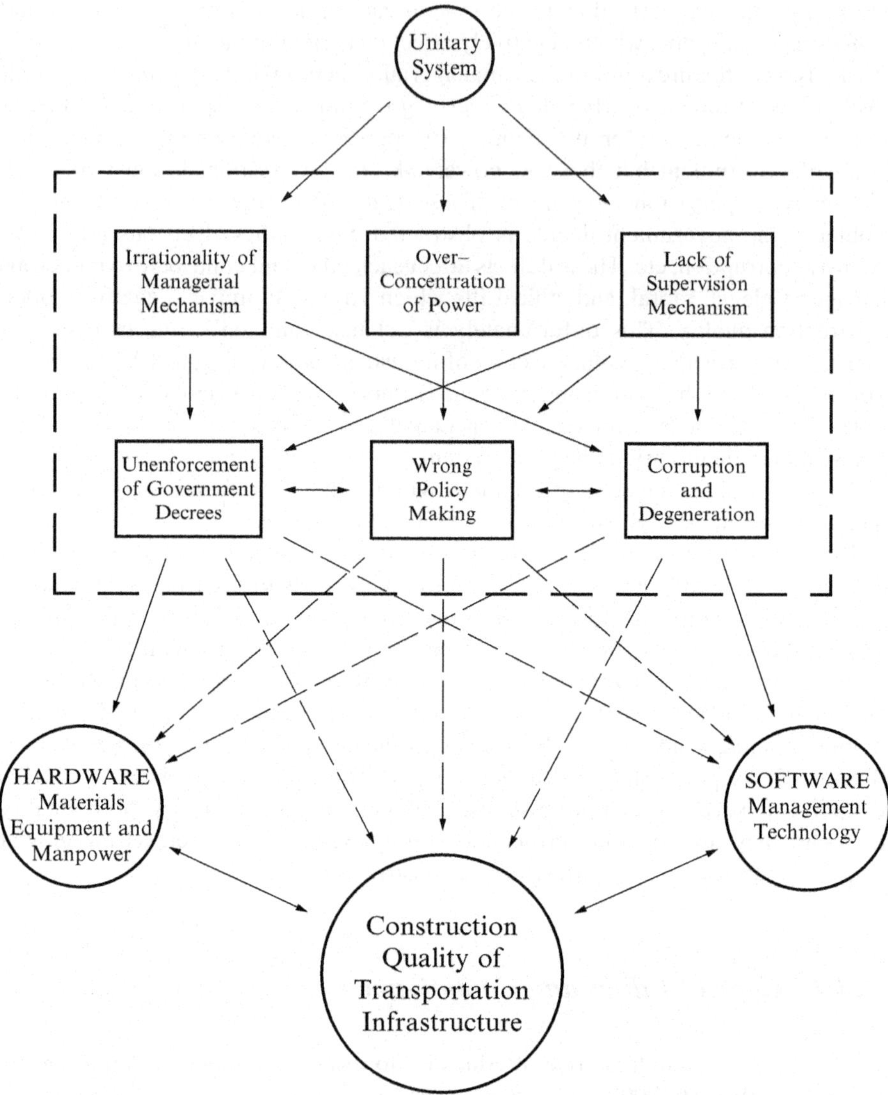

Fig. 10.1 The typical features of unitary managerial system and its negative impacts on construction quality (From Wang 2006)

Many NGOs (civil societies) pursue aspects of anti-corruption. Media are also focusing on the issue and finding new ways to expose it, e.g. the UN Convention against Corruption, which came into force in December 2005; the Inter-American Convention against Corruption which was adopted in 1996; and the World Bank's Integrity Vice Presidency (2001), which was set up to investigate fraud and corruption in Bank-financed projects and staff misconduct.

A low standard of ethics in construction in some countries led the US construction industry to form the Construction Industry Ethics & Compliance Initiative (CIECI 2013), which is open to responsible companies that design and build projects, including architect/engineers, general contractors and subcontractors, to establish a process for the industry to promote integrity and ethical conduct. The goals are the advancement of organizational cultures that encourage and support ethical behavior and compliance with the law, and the sharing of best ethical and compliance practices within the industry.

The Initiative requires each signatory company to pledge to follow six core ethical principles and to participate in an Annual Best Practices Forum to discuss best ethical and business conduct practices among its members and with representatives from government and other organizations. The core principles for each member are to:

(i) Have and adhere to a written Code of Business Conduct. The Code shall establish high ethical values and compliance with the law applicable to the US construction industry;

(ii) Train its personnel as to their personal responsibilities under the Code;

(iii) Work together toward maintaining open competition in the industry, free of conflict of interest and undue influence;

(iv) Have responsibility to each other to share best ethical and compliance practices in implementing the principles;

(v) Participate in the Annual Best Practices Forum;

(vi) Be accountable to the public.

10.5 Ethics Education for Construction Professionals

10.5.1 The Necessity

Current civil and construction engineering and management programs in higher education institutions provide students with sound technical knowledge. However engineers and managers increasingly need to operate in diverse organizational contexts; as such they need the intellectual tools to comprehend the moral diversity within different cultures and different disciplines (Wang et al. 2010). Differences can arise due to varied cultural, religious, environmental, legal and political perspectives, and not surprisingly, can be considerable, affecting an engineer's and a manager's judgment, the public welfare, the environment, and sustainable economic development. Further, in the last few decades, the concept of ethics is no longer limited to an individual's code of conduct; it involves the morally-based practices of groups or disciplines, e.g. a company's corporate responsibilities, and responsibilities and ethical decisions that relate to the environment (Carroll and Buchholtz 2011; Crane and Matten 2007; Elder 2004; Frederick 1994; Lynch 1998). Some ethicists have even split ethics into micro-ethics and macro-ethics (Herkert

2005; Ladd 1980), and to combine these changes or international components in ethics education is a challenge for educators.

Current ethics education curricula, either stand-alone ethics courses or those providing ethics instruction through core courses such as micro-insertion (Davis 2006), do not usually provide students in civil and construction engineering with training on international related ethical practices that reflect the changing world (Herkert 2000).

Ethical issues not only affect individuals, they affect all businesses and construction – both at home and internationally (Berge 2009). Because of a greater likelihood of involvement of construction engineering and management graduates in international projects, there is an imperative for educators to

(i) improve ethics education;
(ii) minimize the negative consequences and impact of unprofessional and unethical behaviors;
(iii) enhance the individual's and organization's competitiveness from a humanistic, individual, and corporate perspective, through ethics education.

As globalization components are incorporated into ethics education of engineers it will allow students to (i) identify the differences in ethical standards and implementation in different countries; (ii) identify the differences in ethical practice and the root causes and the cross impacts of these practices in an international environment; and (iii) apply the principles in making ethical decisions.

It is crucial for educators to pass on the fundamental and critical information regarding the differences in ethics, including universal ethics theory, to their students before they graduate and join the workforce. Interestingly, it has been observed that a notable number of graduate students expressed great interest in ethics issues (Buckeridge 1994; Ehsan and Anwar 2009).

The first step to improve ethics education is to conduct thorough international comparative studies; although there is limited literature available, there are reports of individual (micro-ethics) ethics education practice from different countries (Boni and Berjano 2009; Christensen and Kjolhede 2008; Fotheringham 2008; Frey and O'Neill-Carrillo 2008; Maruyama and Ueno 2007; McGregor 2000). The reform and improvement of ethics education involves basic ethics teaching (Bucciarelli 2008; Buckeridge 1994), methodology improvement (Rabikowska 2009), including the use of modern telecommunication technology (Loui 2005; Weil 2005), and curriculum module development (Davis and Riley 2008; Freeney 2010). The evaluation methods used are an important part of ethics education (Hollander 2005; Sork 2009). Through effective ethics education, students should be able to solve issues in an international context, and differentiate between technical errors and ethical issues. Good case studies are essential.

A comparative study on ethics education and ethics practice will produce new concepts related to global contexts and ethics. Leadership has been listed as one of the components in many universities' missions and ethics is listed as the number one element of the four elements of the "leadership diamond" by Peter Koestenbaum

(Koestenbaum 2002). Krishnasamy raised the issue of "how an ethical perspective can lead to 'Better Humanity'" (Krishnasamy 2008; Stieb 2007).

As corrupt business and construction practices are the norm in some environments, students must learn not only that there are differences in what is considered acceptable practice; they must also learn how to intellectually critique dubious ethical norms. We expect that exploration for and demonstration of universal and/or ubiquitous moral values will be an important component in international ethics education. One of the examples is the concept of Corporate Social Responsibility (CSR) (Baker 2004). CSR focuses on accountability of organizations to a broad group of stakeholders including employees, customers, suppliers, community organizations, subsidiaries and affiliates, joint venture partners, investors and shareholders. CSR policies are based on social justice principles, human rights, and environmental standards. CSR can make our science, engineering and management students better understand that "we do not inherit the earth from our ancestors; we borrow it from our children" – a Native American proverb. Balanced evaluations are necessary.

CSR is important to businesses as it enlightens self-interest because corporations perceived as being socially responsible might gain from getting extra and more satisfied customers. Those perceived as irresponsible may face boycotts. High calibre personnel might be attracted to work for and be committed to responsible firms. Voluntary commitment to social actions and programs might forestall legislation or other government action. When engineers contribute to society, the outcome is a long-term investment in a safer, better-educated, more equitable community, hence firms gain from a more stable business environment. A major principle of CSR is the chain of responsibility. Designers, engineers and producers have a duty to care for human rights and the work practices of communities to ensure the maximum benefits of sustainability.

Comparing codes of ethics from different professional organizations and countries provides an opportunity to identify variations in ethical practices. Although codes of conduct of professional associations help win public confidence and help professions legitimize their control over the professional, the codes do not guarantee compliance by all members due to varied cultural, religious, political, and environmental reasons. Some companies also have formal codes of ethics, key strategic objectives or sets of core values. Some organizations have ethics training programs which teach their workers how to predict and respond to ethical dilemmas that might arise in their jobs.

In addition, overseas professional practice is also informative to study; e.g. the Tokyo Electric Power Company (TEPCO), which had nuclear reactor problems arising after the 2011 tsunami. A later investigation revealed that a senior field engineer, who was dismissed some years prior for whistle-blowing, had alerted TEPCO to problems with their nuclear reactors several years ago (Moret 2011).

Further examples include Apple's 2012 announcement that the Fair Labor Association will start to inspect suppliers, including Foxconn Technology Group's plant in China for poor working conditions for the workers (Perlberg 2012), and the eight-story commercial building that collapsed in Dhaka, Bangladesh on in 2013, killing 1,127 and injuring 2,500 (AP 2013). As a result, the UK Government will push through new legislation to end modern day slavery by forcing companies in the UK to audit their supply chain. The European Commissioner for Trade warned that retailers and the Bangladeshi government could face action from the EU if nothing is done to improve worker conditions. Wal-Mart Stores, J.C. Penney and labor activists are also considering an agreement to improve factory safety in Bangladesh.

Those selecting materials and equipment in design and construction well illustrate that the supply chain issue is now integral in a globalized economy. Supply chains can cause great hardship, and even death for many, especially those living in developing societies. To counter these problems, we need to ask, how do supply chains benefit the poor and disadvantaged? The recently developed Fairtrade concept is now becoming popular in countries such as Australia and the UK. Fairtrade Labeling Organizations (FLO) is a non-profit, multi-stakeholder association (labeling initiatives and producer networks) that develops and reviews Fairtrade Standards and assists certified producers in gaining and maintaining Fairtrade Certification while capitalizing on market opportunities (FLO 2011). Its strategic intent is to work with marginalized producers and workers in order to help them move from a position of vulnerability to security and economic self-sufficiency; to empower producers and workers as stakeholders in their own organizations; and to actively play a wider role in the global arena to achieve greater equity in international trade.

As future scientists and engineers "we want a business that is financially sustainable but not to the extent of using workers that are exploited" (Savaidis 2008). The concept of Fairtrade will raise design and production standards and produce a range of sustainable and ethically manufactured products, which directly profits every aspect of society.

These basic concepts along with others will help students fully understand the importance of ethics and professionalism and apply these principles in their work. It is stated that "an amateur practices until he/she can do something right; a professional practices until he/she cannot do it wrongly" (Lynch 1998), and "professional behavior is usually characterized by making decisions that are in the best interests of the client and not the practitioner" (Kelly et al. 2004). Professional services carry significant moral responsibility and invoke public interest and public good arguments. But some professionals offer their services purely for pecuniary reasons (Low 1996). The process of learning ethics is however not simply confined to the undergraduate curriculum. It should be life-long, for only then can it truly be of major advantage to society. New ethical problems arise continuously, and many do not have easily defined solutions -- they are often surrounded by ambiguities, complexity and ill-defined boundaries. In a global situation, students who enter civil, construction, environmental protection, logistics, and supply chain industries and services, should clearly understand the outcomes that may have economic, legal and social benefits, and outcomes that may have economic, legal and social costs (Buckeridge 2011).

It is clear that there exists an imperative for incorporating globally related ethical components into future ethics education. However, few researchers have dealt with international related ethics topics to date (Feldhaus and Fox 2004; Henderson et al. 2010).

10.5.2 Measures to Enhance Ethics Education

Measures to integrate global components into ethics education for construction majored students and professionals include curricula reform and development, international lecture exchange, study abroad, student exchange, scholar exchange and collaborative research.

Curricula Reform and Development International ethics case studies should play an important role to enhance students' awareness and ability to handle global ethics dilemmas. The case study materials should include typical cases relevant to ethics issues under various international contexts and from different countries. Further, it may also be informative to consider ethical imperatives held by other disciplines, especially when they pertain to how we manage a degrading natural environment (Watts and Buckeridge 2012).

As a first step, a compilation of case studies will be a good reference for student learning, especially when active case studies are discussed and enriched as the project proceeds. The curricula should include the detailed teaching schedule, course content, learning outcomes and objectives, and evaluations.

International Lecture Exchange Lecture exchange can be traced back to 1940s (CARL 2009). For example, in 1947 the Kermit Roosevelt exchange program started with American and British military lecturers exchanging lectures and developing closer relationship between individuals. Exchange lectures have been held every year since 1947. Previously, these lecture exchanges involved scholars physically travelling and visiting each other, but modern technology now permits lectures via virtual collaborative classroom. These lecture exchange programs enhance student learning environments in international related studies (IIE 2009).

Although topics in lecturer exchange focus on the ethical practice and cases in an individual country they also include ethics theory and engineering practice – global aspect; philosophy in engineering and project management; ethics and professional practice in different countries; globalization, ethics and influencing factors; international business ethics; ethics sides of risk management; ethics in construction industry; international professional code of ethics; international engineers registration; ethics in global supply chains; quality, risk, culture, ethics; international corruption – trend, impact; nature and merits of professionalism on international projects; compliance – global considerations.

Study Abroad Traditionally, study abroad programs are commonly seen in the disciplines outside civil and construction engineering and management, such as foreign languages or arts. In the US, study abroad has a long history in higher education institutions associated with language learning (Themudo et al. 2007). In the 1970s and afterwards, shorter study abroad experiences were introduced, standardized for content (Milleret 1990), and both criticized as insufficient for both language improvement and the acquisition of culture (Talburt 1999), as well as lauded for the ability of any length of study abroad as a "major turning point in a student's life" (Cressey 2000). Buckeridge and Brumley (2006) outline the benchmarking of an ethics course offered in Australia, Germany and New Zealand. The project "Ethics for Engineers" has been successfully delivered for well over a decade in each of the three countries. Although it is a peripatetic lecturer, rather than students, that has formed the essence of this model, there has been positive spin-off, with participating students being sufficiently enthused to undertake their practical projects in one of the other countries in the partnership. Modern technology has also provided affordable opportunities for real time engagement between classes in different countries. The course "Ethics for Engineers" provides students with very different cultural and environmental perspectives on engineering, i.e. one Euro-centric, the other a South Pacific perspective. The potential to expand this is significant.

The number of students participating in study abroad programs in the US has increased nearly 150 % in the past decade, to almost 241,800 in 2006 (Fischer 2008). An increasing number of students join programs shorter than 8 weeks instead of traditional semester long study, the most popular over the last decade being in social sciences, business & management, and humanities (IIE 2009).

Previous studies (Wilson 1993) have interpreted the benefits of studying abroad as gaining a: (i) global perspective, including substantive knowledge and perceptual understanding; and (ii) developing self and relationships, as in personal growth and new interpersonal relationships. Participants of study abroad programs "acquire global-mindedness, grow intellectually, and develop personally". Other studies (DeLoach et al. 2003; Kitsantas 2004; Williams 2005) focus on specific issues such as cross-cultural skills, global understanding, culture sensitivity and communication skills. Language acquisition is still among one of the major benefits but often becomes secondary in comparison to professional growth and personal development. Many of these attributes, if not all, are critical to the education and development of construction professionals.

A study abroad program organized by the first author in 2008 (Wang et al. 2010) showed that the students experienced international construction project management and had the opportunity to learn the insights of international project dynamics, which trigged students to start to think about the differences and factors which may affect international construction management. Many of the critical questions were answered during the 3-week tour and afterwards. The students studied construction as a profession in a global context and maintained a competitive edge in the globalization of the construction market.

Student Exchange Student exchange programs are becoming popular in recent years while international travel becomes affordable. Benefits of student exchange include new insights into the culture of the host country; increasing awareness of the traditions, values and expectations in the host country; increasing awareness of the construction industry and processes in the host country; gaining a deeper knowledge and appreciation through their education and extracurricular activities of the implications of global economics on the worldwide construction marketplace (Zilliante and Liska 2007).

Scholar Exchange and Collaborative Research   There are many higher education exchange activities, a prominent example being the Fulbright Program for international educational exchange. The Fulbright program is based on the premise that scholar exchange provides an essential vehicle for mutual understanding between individuals, institutions and future leaders. The US-UK Fulbright Commission was created by treaty in 1948 and since its inception has expanded its program to include grants for study in a wide variety of fields.

10.6   International Registration

It is clear that there is urgent necessity to regulate professional registration to ensure engineers and managers who work in another country or work for a project in another country should have consistent engineering ethics and practice training which meet an international standard.

10.6.1   Profession and Professional Organizations

Professionals are individuals who possess higher learning and competence within a specified discipline (Buckeridge and Brumley 2006). Possession of higher knowledge allows professionals a degree of autonomy (IEAust 2000); this higher knowledge also confers exclusivity, social status, power and not unreasonably, an entitlement to material rewards; but this status is enjoyed in the knowledge that as professionals they must adhere to a code of ethics; it follows that if they digress from these rules, the outcomes will be harsh indeed (Buckeridge 2008). Maister (1997) stated that "professional" is not a label one gives to oneself; it must come from others; however, not all professions possess the same level of recognition by society or industry.

In some countries or regions, professional institutions accept overseas members, accredit overseas programs, have overseas branches, or have "partnerships" with overseas programs. This may lead to the formation of local professional institutions and local professional registration councils. In addition, overseas professional institutions may jointly recognize each other's qualification; e.g. there are six international agreements currently governing mutual recognition of engineering qualifications and professional competence:

Washington Accord (1989): Recognizes substantial equivalence in the accreditation of qualifications in professional engineering, normally of four years duration.
Sydney Accord (2001): Recognizes substantial equivalence in the accreditation of qualifications in engineering technology, normally of three years duration.
Dublin Accord (2002): Accredits substantial equivalence of tertiary qualifications in technician engineering, normally of two years duration.

The greater mobility of professionals is reflected in three international agreements covering equivalence of practicing engineers. This ensures that a person recognized in one country as reaching agreed international standard of competence is only minimally assessed (primarily for local knowledge) prior to obtaining registration in another country that is party to the agreement.

APEC Engineer Agreement (1999): The representative organization in each country creates a "register" of those wishing to be recognised as meeting generic international standards. Other countries should give credit when such an engineer seeks to have his or her competence recognized.

Engineers Mobility Forum (2001): Operates the same competence standard as the APEC Engineer agreement, but any country may join. Parties to agreement are largely engineering bodies.

Engineering Technologist Mobility Forum agreement (2003): A mutual recognition scheme for engineering technologists.

10.6.2 The Need for Improvement

While the development in international professional registration over the last few decades is showing considerable promise, the latter three international agreements (above) are still in the developmental stage; there is still a lack of more specific regulations in particular sectors, trades, and the construction industry. More importantly, any "recognition" is largely based on the equivalence of tertiary education programs. Engineering practice, the examination of ethics, engineering law, and engineering practices are often not covered in the engineering curricula in developing countries and the so-called equivalency of some courses may be little more than political education and deception.

Internationally recognized ethics and professional practice standards and criteria for those who conduct international projects must be mandatory components in all engineering curricula and on-going professional education.

References

All Business. (2006). *Joint venture construction contract award for planned gas-to-liquids project.* Worldwide Projects, Inc. e-document.

Baker, M. (2004). Corporate social responsibility – What does it mean? http://www.mallenbaker. net/csr/definition.php. Accessed 7 July 2014.

Barringer, B. R., & Ireland, R. D. (2006). *Entrepreneurship: Successfully launching new ventures.* Upper Saddle River: Pearson Education.

Berger, H. (2009). Pakistan unrest deterring investments (11/09). http://www.arabianbusiness.com/ 504517-pakistan-unrest-will-discourage-investors. Accessed 17 May 2013.

Bird, S. J. (2003). Ethics as a core competency in science and engineering. *Science and Engineering Ethics, 9,* 443–444.

Boni, A., & Berjano, E. (2009). Ethical learning in higher education: The experience of the Technical University of Valencia. *European Journal of Engineering Education, 34*(2), 205–213.

Bucciarelli, L. L. (2008). Ethics and engineering education. *European Journal of Engineering Education, 33*(2), 141–149.

Buckeridge, J. S. (1994). Introducing philosophy and ethics to the engineering curriculum. *Transactions of the Institution of Professional Engineers New Zealand, 21*(1), 1–4.

Buckeridge, J. S. (2008). Application of ethics in engineering practice… quis custodiet ipsos custodes? In M. Xie & I. Patnaikuni (Eds.), *Innovations in structural engineering and construction*. Proceedings of the fourth international structural engineering & construction conference (ISEC-4), Melbourne, 26–28 Sept 2007. 1, pp. 1273–1278.

Buckeridge, J. S. (2011). *4 Es: Ethics, engineering, economics & environment* (2nd ed.). Sydney: Federation Press.

Buckeridge, J. S., & Brumley, J. (2006). Ethics, engineering and the environment: Is Hawking correct… is it all too late? In G. Rowe & G. Reid (Eds.) *Creativity, challenge, change: Partnerships in engineering education*. Proceedings of the 17th annual AAEE conference, Auckland, 10–13 Dec. 15, pp. 1–8.

CARL. (2009). Kermit Roosevelt exchange lecture collection. Combined Arms Research Library. http://carl.army.mil/resources/archival/kermit.asp. Accessed 12 Dec 2009.

Carroll, A., & Buchholtz, A. (2006). *Business and society: Ethics and stakeholder management*. Mason: Thomson.

Carroll, A. B., & Buchholtz, A. K. (2011). *Business and society: ethics, sustainability, and stakeholder management* (8th ed.). Mason, OH: Cengage Learning.

CFA. (2012). What could the remodeling/construction industry do to improve their reputation as a whole? http://www.buildzoom.com/answers/1017/remodeling-construction-industry-improve-their-reputation. Accessed 18 May 2013.

Christensen, S. H., & Kjolhede, E. E. (2008). Epistemology, ontology and ethics: "Galaxies away from the engineering world"? *European Journal of Engineering Education, 33*, 561–571.

Chua, G. (2013). Sydney Opera House turns 40. *Architecture & Design*, Oct 23, 2013. http://www.architectureanddesign.com.au/news/sydney-opera-house-turns-40. Accessed 7 July 2014.

CIECI. (2013). http://www.ciecinitative.org. Accessed 18 May 2013.

Crane, A., & Matten, D. (2007). *Business ethics: Managing corporate citizenship and sustainability in the age of globalization* (2nd ed.). Oxford: Oxford University Press.

Cressey, W. (2000). Study abroad and area studies. *African Studies, 28*(1), 46–48.

Davis, M. (2006). Integrating ethics into technical courses: Micro-insertion. *Science and Engineering Ethics, 12*, 717–730.

Davis, M., & Riley, K. (2008). Ethics across the graduate engineering curriculum: An experiment in teaching and assessment. *Teaching Ethics, 9*(1), 25–42.

DeLoach, S., Saliba, L., Smith, V., & Tiemann, T. (2003). Developing a global mindset through short-term study abroad: A group discussion approach. *Journal of Teaching in International Business, 15*(1), 37.

Ehsan, N., & Anwar, T. (2009). Professional ethics in construction in Pakistan. *Proceedings of the World Congress on Engineering*. http://www.iaeng.org/publication/WCE2009/WCE2009_pp729-733.pdf.

Elder, K. E. (2004). Ethics education in the consulting engineering environment: Where do we start? *Science and Engineering Ethics, 10*, 325–336.

Elkington, J. (1997). *Cannibals with forks: The triple bottom line of 21st century business*. Oxford: Capstone Publishing.

Engineers Australia (IEAust). (2000). Code of ethics. Engineers Australia. www.engineersaustralia.org.au/ethics. Accessed 11 Feb 2011.

EU. (2009). Chinese construction company starts New York Alexander-Hamilton Bridge renovation projects. *Government Updates, European Union Chamber of Commerce in China*. 16 July 2009.

Fairtrade International (FLO). (2011). http://www.fairtrade.net/. Accessed 7 July 2014.

Feldhaus, C. R., & Fox, P. L. (2004). Effectiveness of an ethics course delivered in traditional and non-traditional formats. *Science and Engineering Ethics, 10*, 389–400.

Fischer, K. (2008). For American students, study-abroad numbers continue to climb, but financial obstacles loom. *The Chronicle of Higher Education, 55*(13), A24–A24.

Fotheringham, H. (2008). *Ethics case studies: Placing ethical practice in an engineering context.* Innovation, Good Practice and Research in Engineering Education – The Higher Education Academy Engineering Subject Centre and the UK Centre for Materials Education EE2008, Liverpool

Frederick, W. (1994). From CSR 1 to CSR2: The maturing of business-and-society thought. *Business and Society, 33*(150), 150–164.

Freeney, S. (2010). *Ethics today in early care and education review, reflection, and the future.* NAEYC: http://www.naeyc.org/files/yc/file/201003/FeeneyWeb0310.pdf, 72–77. Accessed 29 Apr 2015.

Frey, W. J., & O'Neill-Carrillo, E. (2008). Engineering ethics in Puerto Rico: Issues and narratives. *Science and Engineering Ethics, 14*, 417–431.

Haggett, S., & Orr, B. (2013). SNC-Lavalin agrees to 10-year ban from World Bank projects. http://www.cbc.ca/news/canada/montreal/story/2013/04/17/business-snc-lavalin.html. Accessed 16 May 2013.

Henderson, R. (2012). Employment outlook: 2010–2020 Industry employment and output projections to 2020. *Monthly Labor Review, 135*(1), 65–83.

Henderson, R. L., Antelo, A., & St. Clair, N. (2010). Ethics and values in the context of teaching excellence in the changing world of education. *Journal of College Teaching and Learning, 7*(3), 5–11.

Herkert, J. R. (2000). Engineering ethics education in the USA: Content, pedagogy and curriculum. *European Journal of Engineering Education, 25*(4), 303–313.

Herkert, J. R. (2005). Ways of thinking about and teaching ethical problem solving: Microethics and macroethics in engineering. *Science and Engineering Ethics, 11*, 373–385.

Hollander, R. D. (2005). Ethics education at NSF commentary on "standards for evaluating proposals to develop ethics curricula". *Science and Engineering Ethics, 11*, 509–511.

IHS. (2013). Global construction outlook. http://www.ihs.com/products/global-insight/industry-analysis/construction/global.aspx. Accessed 10 May 2013.

IIE. (2009). *Higher education on the move: New development in global mobility.* Washington, DC: Institute of International Education (IIE).

Iriekpen, D. (2009). Nigeria bribe – Halliburton to pay US$559 million settlement. *World News Journal,* Jan 28, 2009. http://africannewsanalysis.blogspot.com/2009/01/nigeria-bribe-halliburton-to-pay-us-559.html. Accessed 16 May 2013.

Kelly, J., Male, S., & Graham, D. (2004). *Value management of construction projects.* Oxford: Blackwell.

Khan, A. H, Azhar, S., & Mahmood, A. (2008). *Quality assurance and control in the construction of infrastructure services in developing countries – A case study of Pakistan.* Proceedings of the first International Conference on Construction in Developing Countries (ICCIDC–I). 4–5 Aug 2008, Karachi, 109–120.

Kitsantas, A. (2004). Study abroad: The role of college students' goals on the development of cross-cultural skills and global understanding. *College Student Journal, 38*(3), 441–452.

Koestenbaum, P. (2002). *Leadership: The inner side of greatness, a philosophy for leaders.* San Francisco: Wiley.

Krishnasamy, S. (2008). Comments on "on 'bettering humanity'" in science and engineering education. *Science and Engineering Ethics, 14*, 291–293.

Ladd, J. (1980). The quest for a code of professional ethics: An intellectual and moral confusion. *AAAS professional ethics project: Professional ethics activities in the scientific and engineering societies* (pp. 154–159). Washington, DC.

Loui, M. C. (2005). Educational technologies and the teaching of ethics in science and engineering. *Science and Engineering Ethics, 11*, 435–446.

Low, L. (1996). *Professionals at the crossroads in Singapore.* Singapore: Times Academic Press.

Lynch, P. (1998). Professionalism and ethics. In P. Sadler (Ed.), *Management consultancy: A handbook of best practice*. London: Kogan Page.

Maister, D. (1997). *True professionalism: the courage to care about your people, your clients, and your career*. New York: Free Press.

Maruyama, Y., & Ueno, T. (2007). *Ethics education for professionals in Japan: A critical review*. Philosophy of Education Society of Australasia, 2007 Conference Presentation, 8 pages.

McGregor, J. (2000). Practice-focused ethics in Australian engineering education. *European Journal of Engineering Education, 25*, 315–324.

Milleret, M. (1990). Evaluation and the summer language program abroad: A review essay. *The Modern Language Journal, 74*(4), 483–488.

Moret, L. (2011). Japan's deadly game of nuclear roulette. *The Japan Times*. http://search. japantimes.co.jp/print/fl20040523x2.html. Accessed 18 May 2013.

Nunnally, S. W. (2007). *Construction methods and management* (7th ed.). Columbus: Pearson Prentice Hall.

Office of Fair Trading (OFT). (2010). Evaluation of the impact of the OFT's investigation into bid rigging in the construction Industry. A report by Europe Economics. http://www.oft.gov. uk/shared_oft/reports/Evaluating-OFTs-work/oft1240.pdf. Accessed 16 May 2013.

Pearl, R., Bowen, P., Makanjee, N., Akintoye, A., & Evans, K. (2005). *Professional ethics in the South African construction industry – a pilot study*. Proceedings of the Queensland University of Technology Research Week International Conference, Brisbane.

Perlberg, H. (2012). Apple says fair labor association will inspect suppliers including Foxconn, *Bloomberg News*. http://www.bloomberg.com/news/2012-02-13. Accessed 18 May 2013.

PRI. (2011). California turns to Chinese company, labor to build most of new Bay Bridge span. http://www.pri.org/stories/world/asia/california-turns-chinese-company-labor-build-bay-bridge-span-6507.html. Accessed 10 May 2013.

Rabikowska, M. (2009). The ethical foundation of critical pedagogy in contemporary academia: (self)-reflection and complicity in the process of teaching. *Pedagogy, Culture and Society, 17*(2), 237–249.

Raftery, J., Pasadilla, B., Chiang, Y., Hui, E., & Tang, B. (1998). Globalization and construction industry development: Implications of recent developments in the construction sector in Asia. *Construction Management and Economics, 16*(6), 729–737.

RICS. (2000). *Guidance notes on professional ethics*. London: Royal Intuition of Chartered Surveyors.

Savaidis, N. (2008). *The age*. April 18, 2008.

Sawyer, J. (2009). Drywall imported from China causing construction problems. Construction Litigation Law Blog on Jan 13, 2009. http://blog.njeifs.com/2009/01/drywall_imported_from_china_ca.html. Accessed 10 May 2013.

Shenkar, O., & Luo, Y. (2004). *International business*. San Francisco: Wiley.

Smith, J. (2010). *Construction management: Understanding and leading an ethical project team*. Santa Monica: Construction Analysis and Planning, LLC.

Sork, T. J. (2009). Applied ethics in adult and continuing education literature. *New Directions for Adult and Continuing Education, 123*, 19–32.

Stieb, J. A. (2007). On "bettering humanity" in science and engineering education. *Science and Engineering Ethics, 13*, 265–273.

Sutherland, J., & Canwell, D. (2004). *Key concepts in management*. London: Palgrave Macmillan.

Talburt, S. (1999). What's the subject of study abroad? Race, gender and "living culture". *The Modern Language Journal, 83*(2), 163–175.

Themudo, D., Page, D., & Benander, R. (2007). Students and faculty perceptions of the impact of study abroad on language acquisition, culture shock, and personal growth. *AURCO Journal Association for University Regional Campuses of Ohio, 13*, 65–79.

TI. (2008). Global corruption report 2008. http://archive.transparency.org/publications/gcr. Accessed 16 May 2013.

US Census Bureau (USCB). (2012). Construction industry statistics. http://www.statisticbrain. com/construction-industry-statistics. Accessed 10 May 2013.

US Securities and Exchange Commission (SEC). (2008). SEC Charges Siemens AG for Engaging in Worldwide Bribery. http://www.sec.gov/news/press/2008/2008-294.htm. Accessed 16 May 2013.

Wang, G. (2006). *Construction quality of transportation infrastructure and managerial system.* The 6th Asia-Pacific transportation development conference. Hong Kong/Macau. 29 May – 1 June 2006.

Wang, G., Lu, H., & Ren, Z. (2010). Globalization in construction engineering and management education. *Journal of Applied Research in Higher Education, 2*(2), 51–62.

Watts, R., & Buckeridge, J. S. (2012). In pursuit of the biological imperative. An intergenerational approach to biological justice. *Biology International, 52,* 14–23.

Weil, V. (2005). Standards for evaluating proposals to develop ethics curricula. *Science and Engineering Ethics, 11,* 501–507.

Williams, T. R. (2005). Exploring the impact of study abroad on students' intercultural communication skills: Adaptability and sensitivity. *Journal of Studies in International Education, 9*(4), 356–371.

Wilson, A. H. (1993). Conversation partners: Helping students gain a global perspective through cross-cultural experiences. *Theory Into Practice, 32*(1), 21.

Zilliante, G., & Liska, R. W. (2007). *Globalization of university construction education: A case study.* Proceedings in the Construction Building Research Conference of the Royal Institute of Chartered Surveyors, Atlanta.

Part II
Implications for Engineering Ethics Education

Chapter 11
Overcoming the Challenges of Teaching Engineering Ethics in an International Context: A U.S. Perspective

Brock E. Barry and Joseph R. Herkert

Abstract The delivery of high-quality instruction in the field of engineering ethics can be challenging. Multiple reports and publications have identified a general hesitation or reluctance among faculty in the engineering disciplines to teach classes in professional ethics. Additional complexity is encountered when delivering that content to an international context. However, a vast array of instructional material is available to assist with the process of preparing to teach engineering ethics. This chapter will identify unique challenges associated with teaching engineering ethics in an international context and it will also provide a detailed assessment of the available teaching resources. The teaching resources evaluated in this chapter will include traditional textbooks, on-line assets, journals, and video-based media. A review of moral development assessment tools will also be provided.

Keywords Engineering • Ethics • Assessment • Instruction • Teaching qualifications • Stage theory • Teaching resources

11.1 Introduction

In the modern world in which today's engineers operate, they have the ability to make an impact (both positive and negative) on society (locally, nationally, and globally) in a manner and to a level never previously considered. For example, advances in biomedical engineering facilitate "enhancement" of human capabilities as well as an unprecedented extension in average life expectancy, while at the same

B.E. Barry (✉)
United States Military Academy, West Point, NY, USA
e-mail: Brock.Barry@usma.edu

J.R. Herkert
Arizona State University, Phoenix, AZ, USA
e-mail: joseph.herkert@asu.edu

© Springer International Publishing Switzerland 2015

C. Murphy et al. (eds.), *Engineering Ethics for a Globalized World*, Philosophy of Engineering and Technology 22, DOI 10.1007/978-3-319-18260-5_11

time advances in weapons engineering make it possible to take lives with precision strikes from great distances. In turn, society reasonably expects those individuals in engineering disciplines to make ethical decisions and to have the moral conviction to adhere to those decisions. Engineering educators have the unique capacity and responsibility to serve as role-models for responsible engineering and, when called upon, to provide instruction in professional ethics.

11.1.1 Framing the Issue

Notions of engineering ethics can vary significantly across the globe, in part due to unique engineering identities in varying national and cultural settings (Downey et al. 2007; Newberry et al. 2011). Newberry et al. suggest that "society conceptions of engineering, and of engineers' professional responsibilities, have evolved differently in different countries and cultures…" (2011). Downey et al. (2007) describe the distinctly different evolution of engineering ethics in France, Germany, and Japan. For example, in France engineers are considered elite civil servants and the topic of engineering ethics receives little attention because ethical practice is simply expected from a person in such a highly respected position. Conversely, in Germany engineers must adhere to a well-established set of guidelines that serves to not only direct their work, but to a greater extent defines their identity as engineers. A traditional strong allegiance to one's employer has resulted in the fairly recent development of individual responsibility and professional ethics in Japan.

Regardless of context, there are several common challenges associated with the process of teaching engineering ethics. Those include comfort with course material and expected qualifications to teach the topic. We view the specific challenges of teaching engineering ethics in an international context as two unique, but not unrelated, perspectives. The first perspective is the one of an instructor, based in the United States, teaching predominately US-born students, and attempting to develop in those students an appreciation for and understanding of ethics in a global context. Conversely, the second perspective is the one of an instructor teaching engineering ethics in a non-US academic institution, attempting to develop the same appreciation for and understanding of ethics in his/her students in their own nation as well as an international context. The solution to both of these challenges lies in the aptitude of the instructor and the resources at the instructor's disposal.

This chapter begins with a discussion of the perceived challenges to teaching engineering ethics in an international context. The challenges discussed include the alleged qualifications to teach engineering ethics and the applicability of universalization. This chapter also provides a detailed review of primarily U.S. resources available to assist the engineering ethics instructor. Each of those resources is assessed relative to its applicability in an international context.

11.1.2 Authors' Lens

It is appropriate to acknowledge the lens through which we, the authors, have prepared this chapter. Both of us were born, raised, and educated in the United States. Dr. Barry spent 10 years working on US and international projects as a civil engineer, prior to entering the field of academics. Dr. Herkert spent more than 5 years working on US projects as an electrical engineer in the power industry, prior to entering the field of academics. Dr. Barry is a licensed professional engineer as was Dr. Herkert for 35 years. Further, we both have experience teaching in traditional engineering programs and teaching engineering ethics courses in the United States and have published individually (Barry and Ohland 2009, 2012; Herkert 1998, 2000, 2001, 2002, 2003, 2005, 2011; Herkert and Banks 2012; Yadav and Barry 2009) and collaboratively (Barry and Herkert 2014) on a variety of topics related to engineering ethics. While we both have routinely taught engineering ethics to international students in our classes, neither of us has spent a significant amount of time teaching engineering ethics as part of course-work outside of the United States. We do make an attempt to routinely incorporate international context in our engineering ethics instruction. Although our knowledge and experience is primarily based on teaching engineering ethics to U.S. students, we argue that much of that knowledge and experience is relevant in a broader, international context.

11.2 Challenges

11.2.1 Qualifications to Teach Engineering Ethics

The code of ethics for the National Society of Professional Engineers states that engineers shall, "perform services only in their area of competence" (Herkert 2000; National Society of Professional Engineers 2011). Similarly, the Engineers Australia code of ethics states that engineers will "act on the basis of adequate knowledge" (Engineers Australia 2010), and the Royal Academy of Engineering in the United Kingdom states that engineers will "perform services only in the areas of current competence" (Royal Academy of Engineering 2011). While the specific qualifications to practice as an engineer are not spelled out in these respective codes, the related society or organization each has a set of well-defined criteria. What then would qualify an engineering educator to be considered "competent" to teach engineering ethics? Without question, there is a certain minimum level of qualification required to teach any particular subject matter. U.S. Institutions with a Carnegie Classification of "DRU," or equivalent, typically require faculty to hold a terminal degree in their field and to demonstrate competence in their area of specialization. A "DRU" classification stands for "Doctoral/Research University" and includes any academic institution that grants a minimum of 20 research doctoral degrees per year (Carnegie Foundation for the Advancement of Teaching 2013). Thus, an individual teaching classical ethics courses in a Department of Philosophy would typically be expected to hold a Ph.D. in philosophy or religion. Likewise, an individual who teaches courses within an engineering discipline would be expected to have a Ph.D. in that subject matter. However, what about the less-well defined area created when the fields of ethics and the engineering disciplines are combined? Is a terminal degree in philosophy and/or engineering required to be qualified to teach a course in engineering ethics? Certainly a myriad of individuals without an engineering, philosophy, or religion background could be identified who are successfully teaching courses in engineering ethics.

A series of reports by the Hastings Center identified a general hesitation to teach professional ethics in a variety of disciplines (Baum 1980; Hastings Center 1980; Kelly 1980; Powers and Vogel 1980; Rosen and Caplan 1980). Arguably, the preparation of faculty to comfortably engage in engineering ethics instruction remains one of the biggest challenges facing engineering ethics education (Herkert 2000). The Hastings Center reports indicate that as of 1980 the qualifications to teach professional ethics within various disciplines were unclear (Baum 1980; Hastings Center 1980; Powers and Vogel 1980; Rosen and Caplan 1980) and more recent literature seems devoid of additional clarity. At the time of publication, the Hastings Center studies revealed that most individuals teaching professional ethics had little or no prior training in that particular subject area. The Hastings Center went so far as to recommend that an individual qualified to teach professional ethics should have an advanced degree in their home discipline (e.g., engineering), as well as a solid background in ethics. However, it should be noted that the Hastings Center does not believe that an advanced degree in moral philosophy or moral theology is required (Rosen and Caplan 1980). In Baum's report (1980) for the Hastings Center, he states that individuals with first-hand field experience are well-suited to teach professional ethics and that qualified individuals should be familiar with the history of their own discipline, including the development of professional societies and codes.

Newberry (2004) states, most "current engineering faculty members are products of the admittedly ethics-deficient undergraduate engineering educational system." That is, most of the individuals we are asking to teach engineering ethics to the next generation of engineers are themselves products of an education with little or no direct discussion of engineering ethics. A possible method for overcoming this lack of knowledge is for the individual to develop an expertise in professional ethics in the course of self-education through reading, use of online resources, discussions with colleagues and, where available, faculty development seminars. A detailed discussion of how faculty seminars and workshops can increase the awareness and comfort of faculty preparing to teach professional ethics is provided by Weil (2003). Thus, while a background and experience in philosophy and engineering might make an individual well-suited to teach engineering ethics, an equally well-suited instructor could come from a background in history of science or technology, technical communications, science and technology studies, etc. Credentials aside, faculty must be enthusiastic about and comfortable with discussing ethical issues and the social implications of engineering (Vogt 2008). Quality instruction can be embodied in the form of the instructional methods and/or the individual responsible for the delivery of the instruction.

11.2.2 Kohlberg, Stage 6 Thinking, and Universalization

Apart from credentials or quality of instruction, the challenge still remains to develop within a student the ability to recognize moral dilemmas, be knowledgeable of professional standards, and apply moral reasoning. That challenge exists both for US and non-US instructors of engineering ethics. To be well-prepared to meet that challenge, it is important to possess a basic understanding of the moral development process. This section begins with an introduction to Lawrence Kohlberg's Stage Theory of Moral Development. Then Stage 6, known as universalization, is identified as the possible impetus for difficulty in understanding ethical issues in an international context.

Kohlberg developed a stage-based theory of moral development that has been in use directly or served as the basis for additional research on moral development for the past 50 years. Kohlberg's original study was conducted with 72 boys from middle and lower class families of Chicago (Crain 2005). The boys were 10, 13, and 16 years of age (Crain 2005). Each child was asked to discuss how they would respond to a series of ethical dilemmas. Kohlberg developed a system to rate each response. From this work Kohlberg asserted that individuals progress through a series of six stages of moral development (see below) (Spodek and Saracho 2006). Kohlberg's theory was in-turn based on Piaget's stage theory of cognitive development.

One of the moral dilemmas developed by Kohlberg and implemented in his research is known as the Heinz Dilemma. This dilemma has been reformulated often, but it is presented below in its original format as prepared by Kohlberg:

> In Europe, a woman was near death from a very bad disease, a special kind of cancer. There was one drug that the doctors thought might save her. It was a form of radium that a druggist in the same town had recently discovered. The drug was expensive to make, but the druggist was charging ten times what the drug cost him to make. He paid $200 for the radium and charged $2,000 for a small dose of the drug. The sick woman's husband, Heinz, went to everyone he knew to borrow the money, but he could get together only about $1,000, which was half of what it cost. He told the druggist that his wife was dying and asked him to sell it cheaper or let him pay later. But the druggist said, "No, I discovered the drug and I'm going to make money from it." Heinz got desperate and broke into the man's store to steal the drug for his wife. (Kohlberg 1981)

Study participants were asked to address a series of questions related to a dilemma, such as the Heinz Dilemma. The questions include: "How should/did the main character act?", "Was it right or wrong?", and "What is the basis of your view?" The questions are designed to probe the reasoning behind an individual's answer (Crain 2005). Answers are not scored based on a correct or incorrect answer, as right or wrong is not part of Kohlberg's scoring system. Rather, they are evaluated with relation to Kohlberg's six stages of moral development. In fact, the specific answer provided during evaluation is unimportant, but the reasoning behind an answer is utilized to assign an individual's thinking to one of the six stages.

Kohlberg's six stages are divided into three levels of morality: preconventional, conventional, and postconventional (Crain 2005; Self and Ellison 1998). In turn, each level contains two stages. Each of the six stages is described in Table 11.1 and further, each stage is presented relative to the Heinz Dilemma. Kohlberg's work was based on the stage theory framework first introduced by Piaget. Piaget's theory is founded on five tenets: qualitatively different stages, each stage is a structured whole, invariant sequence, hierarchical progression, and universalization. Each of the five tenets is described in Table 11.2.

Although Stage 6 was an original part of Kohlberg's theory, late in his life he began to refer to it as a "theoretical stage" that was difficult to identify and evaluate with the available assessment tools (Crain 2005). It is this aspect of moral development that can make it difficult to teach engineering ethics. By far the most common method of engineering ethics instruction is to utilize some sort of scenario, hypothetical dilemma, or real-world example (Yadav and Barry 2009). This method is known as the case-based method of instruction. However, inherent in this system of instruction is the assumption that students have the ability to consider the perspective of multiple individuals associated with the case. As Kohlberg noted, to consider moral dilemmas at this level requires Stage 6 development that is extremely difficult to achieve. While it can be difficult to understand the perspective of a protagonist associated with a particular dilemma that perspective can be increasingly more difficult to comprehend when cultural influences are introduced. There is an expression that says to understand someone else's perspective you need to walk a mile in their shoes. However, it may require more than a mile to truly understand someone's perspective if the shoes they wear are sandals, wooden clogs, or no shoes at all, and you were raised wearing sneakers.

It is only natural to question if Kohlberg's theory of moral development can be equally applied across multiple cultures, and therefore appropriate when considering engineering ethics in an international context.

The final stage in Kohlberg's theory (as well as Piaget's) is universalization. Universalization does not suggest that moral values are universally accepted across cultures. Rather, this stage refers to an individual's ability to take into consideration the viewpoints/perspectives of multiple individuals simultaneously and thus, to be able to reason universally. Each of Kohlberg's and Piaget's individual stages does not refer to specific beliefs, but rather to underlying modes of reasoning (Crain 2005).

Kohlberg and his colleagues tested his theory via cross-sectional and longitudinal studies in a variety of cultures around the world, including Mexico, Taiwan, Turkey, Israel, the Yucatan, Kenya, Bahamas, and India (Crain 2005). For example, Kohlberg stated, "I found no important difference in the development of moral thinking among Catholics, Protestants, Jews, Buddhists, Moslem, and atheists"

Table 11.1 Six stages of Kohlberg's theory of moral development

	Stage level (1–6)	Description	Example with respect to the Heinz Dilemma
Preconventional	Stage 1 – Obedience and punishment orientation stage	An individual assumes that a fixed set of rules have been established by an all-powerful authority. These rules must be obeyed without question	Common Stage 1 responses refer to concern for laws and punishment
	Stage 2 – Individualism and exchange stage	Marked by recognition that different people have different viewpoints. Each person's view is relative to that individual and what is correct is whatever meets the individual's self-interests. At Stage 2, punishment becomes a risk rather than a certainty (as in Stage 1 thinking)	A Stage 2 response to the Heinz Dilemma would focus on what is best for Heinz
Conventional	Stage 3 – Good interpersonal relationship stage	Persons believe that morality is based on both good behavior and interpersonal feelings between individuals	A Stage 3 response to the Heinz Dilemma might be focused on Heinz's responsibility as a husband to help his wife, at whatever cost
	Stage 4 – Maintaining the Social Order Stage	Concepts of Stage 3 are elaborated to society as a whole. Stage 4 individuals place an emphasis on obeying laws, respecting authority, and performing particular duties in an effort to maintain social order	A Stage 4 response to the Heinz Dilemma might express concern for the chaos that would be created in a society where everyone breaks the law
Postconventional	Stage 5 – Social Contract and Individual's Rights Stage	Characterized by individuals who question what makes a good society. The Stage 5 thinker considers society in a theoretical manner and judges a condition on how society ought to operate, rather than how it currently does. An individual at this stage believes that all rational people share two common beliefs. First, there are certain basic rights that are shared by all people and should be protected (e.g., life). Second, unfair rules or laws can be changed to improve society through democratic procedures	A Stage 5 respondent to the Heinz Dilemma may make a strong statement that laws are social contracts that should not be broken. However, the Stage 5 respondent would view the wife's right to life as justification for Heinz's actions
	Stage 6 – Universal principles stage	Initially defined this stage to deal with the fact that the Stage 5 democratic process for addressing disputes may not always produce justice. Just decisions can be reached by evaluating a dilemma through the eyes of all the parties involved	Stage 6 thinking would consider the viewpoint of the druggist, Heinz, and Heinz's wife

Table 11.2 Five tenets of Piaget's stage theory

Tenets	Description
Qualitatively different stages	Qualitatively different ways of thinking. Each stage presents a unique perspective for an individual to evaluate how they would respond to a particular moral dilemma
Each stage a structured whole	Each stage represents a pattern of thought that can be identified within various issues. While it is most common that an individual's thinking be contained entirely within a particular stage (e.g., Stage 2 or 3), it is possible and likely for an individual to exhibit thinking associated with two adjacent stages (e.g., Stage 2 and 3). An individual that show signs of thinking in two stages would be considered in transition to the higher stage
Invariant sequence	Progress through the stages occurs in an invariant sequence and skipping of discrete stages does not occur. Each stage is a prerequisite for the subsequently higher stage
Hierarchical progression	Development is hierarchical in form. Each stage requires the individual to understand and to integrate the insight gained at earlier stages into a new framework. The concepts valued in a lower stage are all incorporated in the more advanced principles of subsequent stages. For example, an individual with Stage 3 thinking can still understand Stage 2 arguments, but this individual now uses a more cognitively adequate form of evaluating moral dilemmas
Universalization	The stage concept is universal across all cultures

(Kohlberg 1981). Kohlberg would suggest that social experiences have a greater influence on moral development than the influence of cultural factors.

It should be acknowledged that like many significant social theories, Kohlberg's theory has received a fair amount of criticism. Kohlberg welcomed the dialogue with his critics. He believed that without conflict and dialogue, we cease to develop (Palmer et al. 2001). Much of the criticism of Kohlberg's work tends to focus on his original theory and the research conducted to develop his doctoral dissertation. In particular, many of his critics note that the participants involved with his original research were all adolescent, male, middle-class individuals from a particular American city. Yet, Kohlberg's claim that the theory is universal would suggest that it can be generalized without concern for gender, age, and culture. The subject literature indicates that the leading critic of Kohlberg was Carol Gilligan (Crain 2005; Jorgensen 2006; Kohlberg et al. 1983; Kurtines and Gewirtz 1984). Gilligan believed that males use a gender-specific approach to solve problems of interpersonal conflict (Gilligan 1982). It is interesting to note that at no time did Kohlberg suggest that males have a more developed sense of justice than females. However, he did believe that women, as a subset of any culture, need to be provided "the experience of participation in society's complex, secondary institutions through education and complex work responsibility" (Kohlberg et al. 1983). In accordance with longitudinal studies that illustrate a need for intellectual stimulation to promote stage advancement, Kohlberg is in fact advocating for the involvement of women in society as a means of promoting their moral development. In Kohlberg's words, "...if women were not provided with the experience of participation in society's complex, secondary institutions through education and complex work responsibility, then they were not likely to acquire those societal role-taking abilities necessary for development..." (Kohlberg et al. 1983).

Although much of the literature appears to sensationalize the shared criticism between Kohlberg and Gilligan, it should not be overlooked that the two often collaborated on research and publications. Jorgensen (2006) performs an in depth evaluation of the supposed rift between Kohlberg and Gilligan. He uses a personal interview with Gilligan and a posthumous study of Kohlberg's writings to provide evidence that the two individuals were actually supportive of each other, rather than complete adversaries. When asked if Gilligan saw herself as a critic of Kohlberg, she responded, "I never saw myself as a critic, no" (Jorgensen 2006). In fact it appears that Gilligan expresses great admiration for Kohlberg's work. Likewise, Kohlberg saw the work of Gilligan as an enlargement of his own theory, not an attempt to falsify his work. Thus, it would appear that although many texts paint Gilligan and Kohlberg as adversaries, neither Kohlberg nor Gilligan truly saw Gilligan as a critic of Kohlberg's theory. Both individuals simply viewed Gilligan's work as an expansion of Kohlberg's theory (Jorgensen 2006).

In closing this section, we wish to emphasize that the intent of introducing Kohlberg's stage theory, in particular Stage 6 – Universalization, was not to suggest that it is particularly well suited for teaching engineering ethics in an international context. Rather, the concept of universalization it presented as a potential explanation as to why all individuals find it difficult to identify and evaluate moral dilemmas in cultures beyond their own.

11.3 Resources

11.3.1 A General Research Question

In the prior section we made the case that there is an apparent apprehension among most engineering faculty to teach course content related to ethics. In addition, we noted that based on Kohlberg's stage theory of moral development, the expectation that students can understand the perspective of various individuals associated with a moral dilemma, may in fact be too aspirational. While Kohlberg's theory has been demonstrated as equally applicable across cultures, it is only reasonable to expect that students would have an increased difficulty in understanding a different perspective when the dilemma is set in a culture they are unfamiliar with.

We were accordingly motivated to ask the question, "what resources exist to assist with the instruction of engineering ethics concepts in an international context?" As noted in the introduction to this chapter, we consider an "international context" to include both instruction to US students about engineering ethics in non-US countries and instruction to non-US students about engineering ethics in other countries. With this in mind, we conducted a general investigation of available resources (primarily in the U.S.). While we utilized common education-research methods in the process of investigating this research question, we do not claim that the results are all encompassing. It is entirely possible that some resources were overlooked. We have limited the identified resources to be inclusive of only those items that an instructor would likely utilize in a classroom or pedagogical strategies. It is also appropriate to note that the education resources associated with engineering ethics are continuously evolving. The resources identified herein represent a "snapshot" of the content in the 2011–2013 timeframe. The resources we have identified and will discuss can be grouped into specific categories: textbooks, websites, video, journals, and assessment tools. Each of these categories is discussed separately.

11.3.2 Textbooks

Whether in traditional hard copy or the increasingly popular electronic format, textbooks remain the primary resource used in the instruction of engineering ethics. We were personally familiar with many of the texts commonly in use (for example see Baura (2006), Fleddermann (2008), Gunn and Vesilind (2003), Harris et al. (2009), Herkert (2000), and Martin and Schinzinger (1996)). However, to be more encompassing, we implemented a process of contacting all of the major textbook publishing companies to address a series of questions. Those questions included: "What texts do you currently sell related to engineering ethics?," "Can you share the volume of each title sold?," "What volumes are sold outside of the US?," and "Is the textbook translated into languages beyond English?" Table 11.3 summarizes the results of that inquiry process. Only those publishers that responded to the request for information are listed in this table.

It was not surprising that the majority of publishers were either unable or unwilling to share data related to the annual number of volumes sold and thus, their share of the market. It was simultaneously encouraging that the McGraw-Hill texts were published in multiple languages, but also disappointing to realize that other publishers did not report similar information.

Anecdotally, there are a number of widely used textbooks that relate to engineering ethics. However, the "most popular" or most widely used textbook was not determined through this inquiry.

We performed an inspection of all of the textbooks listed in Table 11.3 and noted that nearly all of them are very good at introducing ethical frameworks and discussing the need to consider ethical dilemmas from various perspectives (Kohlberg's Stage 6). Several of these texts refer to cases that occur in international locations, but then fall short of actually discussing the implications or challenges that arise when ethical frameworks are applied in an international context foreign to the reader. The texts by Harris et al. (2009), Gunn and Vesilind (2003), and Robinson et al. (2007) each dedicate a portion of their content to the specific discussion of engineering ethics in an international context. For example, Harris et. al (who are from the U.S.) include a chapter titled "International Engineering Professionalism"

Table 11.3 Engineering ethics textbook data reported by publishers

Publisher	Titles	Volume		Translations
		US	Outside US	
John Wiley & Sons, Inc.	Speight and Foote (2011) – *ethics in science and engineering*	Not provided	Not provided	None reported
	van de Poel and Royakkers (2011) – *ethics, technology, and engineering*			
	Tavani (2011) – *ethics & technology*			
McGraw – Hill	Martin et al. (2010) – *introduction to engineering ethics*	2,500/year	Not provided	Chinese (long and short form)
	Martin and Schinzinger (2005) – *ethics in engineering*	2,200/year		Korean
				Serbian
Oxford University Press	Seebauer and Barry (2001) – *fundamentals of ethics*	Not provided	Not provided	None reported
	Dunwoody (2006) – *fundamental competencies for engineers*			
Elsevier Academic Press	Baura (2006) – *Engineering ethics: an industrial perspective*	Not provided	Not provided	None reported
	Vallero (2007) – *biomedical ethics for engineers*			
	Robinson et al. (2007) – *engineering, business and professional ethics*			
Pearson Education	Fleddermann (2008) – *engineering ethics*	Not provided	Not provided	None reported
Cengage Wadsworth Thomson Cole	Harris et al. (2009) – *engineering ethics*	Not provided	Not provided	None reported
	Gunn and Vesilind (2003) – *hold paramount*			

and discuss common international issues such as bribery, extortion, nepotism, and exploitation. Gunn & Vesilind dedicate an entire chapter to what they refer to as "overseas work" (perhaps assuming that the majority of their textbooks would be sold in developed countries) and discuss such things as environmental racism, human rights, and cross-border politics. Notably, Gunn is a resident of New Zealand. Robinson et al. (all residents of the United Kingdom) discuss globalization, human rights, child labor, and non-governmental agencies. While each textbook we reviewed had its particular strengths, the three textbooks by Harris, et al., Gunn and Vesilind, and Robinson et al. make a clear case for inclusion of ethics in an international context in the education of engineers.

Finally, the edited volume of which this chapter is a part can also be listed among the books that contribute to a better understanding of the issues related to engineering ethics in the modern society and in particular in an international context.

11.3.3 Websites

Reportedly, just over 1/3 of the world's population has access to the internet (Internet World Stats 2012). Naturally, it would be reasonable to expect that individuals seeking an engineering degree at technology enabled academic institutions throughout the world would have an even higher percentage of access. With such readily available information, websites and the information contained within them are a viable resource for engineering ethics instruction. To determine the extent of those resources, we executed a search for U.S. based websites in a manner similar to what would be expected of an engineering ethics instructor and utilized a basic internet search engine. Each identified website was evaluated relative to its stated mission, nature of its content, and availability of international content. That information is summarized in Table 11.4. The websites identified herein are considered to be the most relevant and contain the most useful information. A search specific to individual countries was not performed.

The Online Ethics Center (OEC) for Engineering and Research (www. onlineethics.org) (National Academy of Engineering 2013), was initially funded through a series of NSF grants. The site is currently maintained by the Center for Engineering, Ethics and Society of the National Academy of Engineering. The mission of the OEC, as stated on the website, is "to provide engineers and engineering students with resources for understanding and addressing ethically significant problems that arise in their work, and to serve those who are promoting learning and advancing the understanding of responsible research and practice of engineering" (National Academy of Engineering 2011). An extensive number of cases and scenarios, as well as related discussion points can be found on the site. Useful from an international context perspective, the site contains codes of ethics for engineering societies from multiple countries and published in multiple languages. Finally, the OEC website includes essays that cover international ethics issues such as free trade agreements, professional organizations, and bribery.

The University of Illinois has recently developed an NSF funded National Center for Professional and Research Ethics (NCPRE) (www.nationalethicsresourcecenter. net) (National Ethics Resource Center 2012). The mission of the NCPRE is to "develop, gather, preserve, and provide comprehensive access to resources related to ethics for teachers, students, researchers, administrators, and other audiences" (Gudeman 2011). The well-organized NCPRE website, Ethics CORE, contains individual course lectures, course syllabi, role-playing scenarios, and various discussion

Table 11.4 Summary of comprehensive engineering ethics websites

Organization	URL	Mission	Website content	International content
Online Ethics Center for Engineering and Research (National Academy of Engineering)	www.onlineethics.org	"…to provide engineers and engineering students with resources…"	Cases, scenarios, discussion points, codes	Multiple codes of ethics from international locations and published in multiple languages, essays on international engineering ethics topics
National Ethics Resource Center (University of Illinois)	www.nationalethicsre-source center.net	"…develop, gather, preserve, and provide com-prehensive access to resources…"	Course lectures, syllabi, cases, scenarios, codes	Some international codes of ethics
National Institute for Engineering Ethics (Texas Tech University)	www.niee.org	"…promoting the study and application of ethics in engineering education…"	Cases, complete courses, codes	No direct international content
Engineering ethics at Texas A&M	http://ethics.tamu.edu	Not stated	Cases, essays, conference proceed-ings	Some essays with international topics

scenarios. The site also includes the code of ethics for ninety-nine different societies, organizations, or companies; many from international locations.

The National Institute for Engineering Ethics (NIEE) is currently organized as an institute at Texas Tech University (TTU) within TTU's Murdough Center for Engineering Professionalism. The NIEE's mission is to promote the study and application of ethics in engineering education and throughout the profession of engineering. NIEE provides a series of professional ethics workshops, seminars, presentations, and distance learning opportunities. The NIEE website (www.niee.org) also contains a significant number of cases and modules designed for classroom use. In addition, NIEE obtained financing for, produced, and markets a series of highly successful engineering ethics videos (National Institute for Engineering Ethics 2013). These videos are each discussed in the subsequent section of this chapter.

The ethics webpage maintained by the Department of Civil Engineering at Texas A&M University (http://ethics.tamu.edu) was one of the first repositories for engineering ethics case studies and essays (Texas A&M University 2013). This site also contains proceedings from engineering ethics workshops and conferences, as well as links to other engineering ethics sites. This website's strength remains the quality of case studies and essays available on the site; many of which are set in an international context and provide guidance for use in a classroom. For example, there are detailed case studies on the design of a farming plow to assist peasant farmers in Mexico and an (actual) aircraft accident in the United Kingdom.

11.3.4 Videos

Video provides the viewer with a unique visual experience in considering different actions and ethical perspectives of the main characters. Thus, if produced well, engineering ethics videos can be a highly effective instructional medium to allow students to view dilemmas in an international context.

The earliest engineering ethics video to experience extensive use in the U.S. was titled, *Gilbane Gold* © (1989). This video was produced in cooperation with the National Society of Professional Engineers, but is now distributed solely by the National Institute for Engineering Ethics (NIEE). The 24-min storyline involves a computer component manufacturing company discharging lead and arsenic into a municipal sewer system. The environmental engineer for the company wrestles with the ethical implications of being pressured to increase discharge rates that meet the letter of the law, but exceed limits intended by the spirit of the law. The video addresses the concepts of protection of public health and the environment, quality of life and the welfare of individuals, free enterprise, and personal integrity. While the concepts addressed in *Gilbane Gold* are easily transferable to an international context, the film is set entirely in the United States, is presented in English, and has not been subtitled in other languages.

Incident at Morales© (2003) was the first DVD of its kind to focus on the unique aspects of professional ethics in an international context. The video was produced and is distributed by the NIEE. The video has a 36 min run time and is partially set in the fictitious town of Morales, Mexico. The script focuses on the ethical issues faced by a multinational company (a U.S. based manufacturing company recently bought out by a French chemical conglomerate) that seeks to rapidly build a production facility in Mexico to gain a competitive edge over their competitors. The plot provides viewers with opportunities to consider ethical decision making, the role of a code of ethics in the decision making process, the applicability of a code of ethics in an international context, and the obligations of engineers beyond fulfillment of a contract with a client or customer. The video has been subtitled in 12 languages, and can be used in concert with a study guide or other available educational content. The study guide provides discussion questions, including a series of questions related to cultural issues in an international context. Upon initial release, the NIEE distributed complimentary copies of this video to 341 U.S. academic institutions and 280 international academic institutions.

A video titled *Henry's Daughters*© (2010) is the most recent educational DVD released by the NIEE. The plot of this DVD focuses on a retired, but well-connected, professional engineer and lobbyist, as well as his two daughters. One daughter is a professional engineer employed by a state-level Department of Transportation, while the second daughter is a recent college graduate and intern with a startup "smart" automotive company. Henry and each of his daughters play specific, but in many cases conflicting roles in the competition for establishment of a smart highway system. The multiple layers of the storyline introduce the audience to ethical considerations, including a licensed professional engineer's obligation to society – not just to fulfilling a contract with a client or customer, technical and non-technical solutions to ethical dilemmas, and ethical implications of workplace issues, such as gender discrimination. The 32 min run time makes it well suited for a typical in-class viewing, followed by discussion. An accompanying 24-page study guide includes a review of the storyline, a list of characters, and suggested assignments. The DVD is embedded with subtitles in 12 languages. The NIEE has distributed complementary copies of *Henry's Daughters* to 350 colleges and universities in the United States and approximately 200 academic institutions beyond the United States. While the ethical issues are easily transferred to an international context, the story is set entirely within the United States.

The American Society of Civil Engineers, in collaboration with 11 other partnering organizations (including the World Bank, the World Federation of Engineering Organizations, and the World Justice Project) produced a video titled *Ethicana* © (2009). This 42-min film is set in the fictitious county of Ethicana and depicts corruption in the engineering and construction industry. The film also addresses issues of bribery, political greed, cultural bigotry, and moral courage. The video is available with subtitles in Arabic, Bulgarian, Chinese Simplified, Chinese Traditional, Czech, Danish, Dutch, Farsi, French, German, Greek, Hindi, Hungarian, Indonesian, Italian, Japanese, Korean, Malay, Norwegian, Polish, Portuguese, Russian, Spanish, Swedish, Thai, Turkish, Urdu, and Vietnamese. Without question, among the education videos widely available in the U.S., Ethicana is the best suited video to assist with the instruction of engineering ethics in an international context.

11.3.5 *Journals*

Professional journals are an often overlooked, but highly valuable resource for instructors charged with the delivery of engineering ethics content. Some of the best journals with engineering ethics research and pedagogy content include: Springer's *Science and Engineering Ethics*, *IEEE Technology and Society Magazine*, and the Society for Ethics Across the Curriculum's *Teaching Ethics*. Further, several other journals routinely publish articles on engineering ethics research and instruction, including the American Society for Engineering Education's *Journal of Engineering Education*, the *International Journal of Engineering Education* (published by TEMPUS) and the American Society of Civil Engineers' *Journal of Professional Issues in Engineering Education and Professional Practice*. Two of the more illustrative articles useful for either self-education or direct discussion in a classroom are:

Engineering Ethics and Identity: Emerging Initiatives in Comparative Perspectives (Downey et al. 2007) and *Internationalizing Professional Codes in Engineering* (Harris 2004).

11.3.6 Assessment of Understanding

The primary objective of an instructor is to facilitate learning. Without formal assessment, it is difficult if not impossible to confirm that learning has taken place. Course specific assessment tools can be developed to confirm that students can achieve at the lower levels of Bloom's Taxonomy of Cognitive Development (Bloom et al. 1984). For example, it is not complicated to design a valid and reliable assessment to determine if a student can define Kantianism or to identify the moral dilemmas in a scenario. However, to truly assess a student's ability to reach the higher levels of cognitive development, such as analysis, synthesis, and evaluation, requires a more complex assessment of moral reasoning.

Kohlberg developed a moral reasoning assessment tool known as the Moral Judgment Interview (MJI). In the application of this tool, participants read three prepared moral dilemmas (including the previously discussed Heinz dilemma). Participants then provide oral responses to a series of standardized questions that the test administrator uses to probe the participant's reasons for their statements. The focus of the assessment is the justification for the participant's reasoning, rather than merely right versus wrong. Scores are based on the relation between their responses to the dilemmas and Kohlberg's six pre-defined stages. The MJI has been shown to be a valid and reliable assessment of moral reasoning (Colby and Kohlberg 1987; Kohlberg 1981). Other general moral reasoning assessment tools have been developed by various researchers, but they each grounded their studies in the works of Kohlberg and in-turn Piaget.

While the MJI was the first well know tool for assessment of moral reasoning, the most commonly used tool today is the Defining Issue Test (DIT) (Killen and Smetana 2006) as developed by James Rest (Self and Ellison 1998). The DIT is a paper-and-pencil-based multiple-choice examination that can be easily administered to large groups and can be quickly computer scored. The relative ease associated with administration and scoring of the DIT is one reason that it has essentially replaced the MJI. In taking the DIT, participants read five moral dilemmas, and then make a selection from a 3-point scale relative to what they believe the protagonist in the dilemmas should do. Then participants evaluate twelve pre-written items to identify which items they believe to be the most important in addressing the dilemma. Finally, participants rank each of the twelve items using a 5-point Likert-type scale (Rest 1994, 1999). DIT assessments result in a P-score, which can be defined as the relative importance that a subject gives to items that represent moral thinking (Duckett et al. 1997; Rest 1999). The most recent version of the Defining Issues Test is known as the DIT-2. In the second version, the instructions have been revised and the dilemmas have been updated (Rest 1999). Both the DIT and the DIT-

2 have been shown to be valid and reliable measures of moral reasoning (Duckett et al. 1997; Duckett and Ryden 1994; Self and Ellison 1998; Sutton 1992).

Research has been conducted that utilizes the MJI, DIT, and similar instruments in the study of applied ethics within various professions (Bebeau 1994, 2002; Drake et al. 2005; Duckett et al. 1997; Duckett and Ryden 1994; Rest 1994; Self and Ellison 1998; Self et al. 1989). Each of these assessment tools has proven to be valid for the evaluation of general moral reasoning, but they have further been shown to lack the sensitivity and context to capture the unique aspects of professional ethics within specific disciplines. Thus, a number of discipline-specific moral reasoning assessment tools have been developed. Barry and Ohland (2009) discuss several assessment tools for the health professions, business, and law.

Borenstein, Drake, Kirkman, and Swann have published a journal article providing results for a discipline specific moral reasoning assessment tool they call the Engineering and Science Issues Test (ESIT) (2010). The ESIT is patterned after the DIT-2, but uses six scenarios in the context of science and engineering (Borenstein et al. 2005).

Canary et al. (2012) have used a shortened version of the ESIT to evaluate three instructional models (traditional standalone course, hybrid online/face-to face course, ethics material embedded in a required course) for delivering ethics instruction to graduate students in engineering and science. One of the findings from that study was that non-native English speakers did not fare as well on the ESIT as their English speaking counterparts. Specifically they state that the results "indicate that caution should be used when using these measures in non-native English speaking samples" (Canary et al. 2012). Further, they state that "these measures might reflect cultural knowledge that foreign students studying in the United States do not have" and "...these differences also might reflect cultural differences in ethical interpretations" (Canary et al. 2012). Finally, they also point out that yet another reason for the disparity could be as simple as the language used to describe the scenarios and response choices may be difficult for non-native English speakers to understand.

While the MJI, DIT, DIT-2, are well designed and proven measures of moral reasoning, only the ESIT places dilemmas in the context of the engineering discipline. The ESIT is still in the early stages of application, but it has the potential to be a highly valued engineering ethics research tool. However, there has been limited evaluation of the ESIT for cultural bias (Canary et al. 2012) and the ESIT was not specifically designed for assessment of moral reasoning in an international context.

11.4 Summary

Various reports have indicated that there is a general hesitancy among engineering faculty to teach course content related to engineering ethics. The perceived challenge could be based in part on concern for credentials, lack of content knowledge, and/or lack of available resources. Those challenges are only magnified when instructors are asked to present course content to an international audience or to present materials based in an international context. The reality is that excellent resources currently exist in the form or textbooks, websites, videos, journals, and moral development assessment tools (while we have focused primarily on U.S. resources, educators in a number of other regions including Japan and Western Europe have also been actively engaged in the field). Unfortunately, the vast majority of each of these resource types was prepared with only a western context in mind. As the world continues to see growth in the global reach and influence of engineering, there is a clear need for more instructional resources that place engineering ethics in an international context. This anthology makes a welcome contribution.

References

American Society of Civil Engineers (Producer). (2009). Ethicana.

Barry, B. E., & Herkert, J. R. (2014). Engineering ethics. In A. Johri & B. Olds (Eds.), *Cambridge handbook of engineering education research*. New York: Cambridge University Press.

Barry, B. E., & Ohland, M. W. (2009). Applied ethics in the engineering, health, business, and law professions: A comparison. *Journal of Engineering Education, 98*(4), 377–388.

Barry, B. E., & Ohland, M. W. (2012). ABET Criterion 3.f: How much curriculum content is enough? *Science and Engineering Ethics, 18*(2), 369–392.

Baum, R. J. (1980). *Ethics and engineering curricula*. Hastings-on-Hudson, The Hastings Center, Institute of Society, Ethics, and the Life Sciences.

Baura, G. D. (2006). *Engineering ethics: An industrial perspective*. Boston: Elsevier Academic Press.

Bebeau, M. J. (1994). Influencing the moral dimensions of dental practice. In J. R. Rest & D. Narvâaez (Eds.), *Moral development in the professions: Psychology and applied ethics* (pp. 121–146). Hillsdale: L. Erlbaum Associates.

Bebeau, M. J. (2002). The defining issues test and the four component model: Contributions to professional education. *Journal of Moral Education, 31*(3), 271–295.

Bloom, B. S., Krathwohl, D. R., & Masia, B. B. (1984). *Taxonomy of educational objectives: The classification of educational goals*. New York: Longman.

Borenstein, J., Kirkman, R., & Swann, J. (2005). Engineering and Science Issues Test (ESIT) *Qeustionaire* (1.0p ed.): Georgia Institute of Technology.

Borenstein, J., Drake, M., Kirkman, R., & Swann, J. (2010). The Engineering and Science Issues Test (ESIT): A discipline-specific approach to assessing moral judgment. *Science and Engineering Ethics, 16*(2), 387–407.

Canary, H., Herkert, J. R., Ellison, K., & Wetmore, J. (2012, June 2012). *Microethics and macroethics in graduate education for scientists and engineers: developing and assessing instructional models*. Paper included in the proceedings for the American Society for Engineering Education Annual Conference & Exposition, San Antonio, Texas.

Carnegie Foundation for the Advancement of Teaching. (2013). The Carnegie classification of institutions of higher education. Retrieved 2 Aug 2013, from http://classifications. carnegiefoundation.org/descriptions/basic.php

Colby, A., & Kohlberg, L. (1987). *The measurement of moral judgment*. Cambridge/New York: Cambridge University Press.

Crain, W. C. (2005). *Theories of development: Concepts and applications* (5th ed.). Upper Saddle River: Pearson Prentice Hall.

Downey, G., Lucena, J., & Mitcham, C. (2007). Engineering ethics and identity: Emerging initiatives in comparative perspective. *Science and Engineering Ethics, 13*(4), 463–487.

Drake, M. J., Griffin, P. M., Kirkman, R., & Swann, J. L. (2005). Engineering ethical curricula: Assessment and comparison of two approaches. *Journal of Engineering Education, 94*(2), 223–231.

Duckett, L. J., & Ryden, M. B. (1994). Education for ethical nursing practice *Moral development in the professions: Psychology and applied ethics* (pp. xii, 233 p.). Hillsdale: L. Erlbaum Associates.

Duckett, L. J., Rowan, M., Ryden, M., Krichbaum, K., Miller, M., Wainwright, H., & Savik, K. (1997). Progress in the moral reasoning of baccalaureate nursing students between program entry and exit. *Nursing Research, 46*(4), 222–229.

Dunwoody, B. (2006). *Fundamental competencies for engineers.* Don Mills: Oxford University Press.

Engineers Australia. (2010). Code of ethics. Retrieved 11 Mar 2013, from http://www. engineersaustralia.org.au/sites/default/files/shado/About%20Us/Overview/Governance/ codeofethics2010.pdf

Fleddermann, C. B. (2008). *Engineering ethics* (2nd ed.). Upper Saddle River: Pearson Education.

Gilligan, C. (1982). *In a different voice: Psychological theory and women's development.* Cambridge, MA: Harvard University Press.

Gudeman, K. (2011). Illinois to develop a national center for ethics in science, mathematics, and engineering. Retrieved 3 Aug 2011, from http://www.ece.illinois.edu/mediacenter/article.asp? id=1162

Gunn, A. S., & Vesilind, P. A. (2003). *Hold paramount: The engineer's responsibility to society.* Pacific Grove: Thomson-Brooks/Cole.

Harris, C. (2004). Internationalizing professional codes in engineering. *Science and Engineering Ethics, 10*(3), 503–521. doi:10.1007/s11948-004-0008-6.

Harris, C., Pritchard, M., & Rabins, M. (2009). *Engineering ethics: Concepts and cases* (4th ed.). Belmont: Wadsworth Cengage Learning.

Hastings Center. (1980). *The teaching of ethics in higher education: a report.* Hastings-on-Hudson, Hastings Center, Institute for Society, Ethics, and the Life Sciences.

Herkert, J. R. (1998). Sustainable development, engineering and multinational corporations: Ethical and public policy implications. *Science and Engineering Ethics, 4*(3), 333–346.

Herkert, J. R. (2000). *Social, ethical, and policy implications of engineering: Selected readings.* New York: IEEE Press.

Herkert, J. R. (2001). Future directions in engineering ethics research: Microethics, macroethics and the role of professional societies. *Science and Engineering Ethics, 7*(3), 403–414.

Herkert, J. R. (2002). Continuing and emerging issues in engineering ethics education. *The Bridge, 32*(3), 8–13.

Herkert, J. R. (2003). Professional societies, microethics, and macroethics: Product liability as an ethical issue in engineering design. *International Journal of Engineering Education, 19*(1), 163–167.

Herkert, J. R. (2005). Ways of thinking about and teaching ethical problem solving: Microethics and macroethics in engineering. *Science and Engineering Ethics, 11*(3), 373–385. doi:10.1007/s11948-005-0006-3.

Herkert, J. R. (2011). Ethical challenges of emerging technologies. In G. E. Marchant, B. R. Allenby, & J. R. Herkert (Eds.), *The growing gap between emerging technologies and legal-ethical oversight: The pacing problem* (Vol. 7, pp. 35–44). New York: Springer.

Herkert, J. R., & Banks, D. A. (2012). I have seen the future! ethics, progress, and the grand challenges for engineering. *International Journal of Engineering, Social Justice, and Peace, 1*(2), 109–122.

Internet World Stats. (2012). Usage and population statistics. Retrieved 12 Mar 2013, from http:// www.internetworldstats.com/stats.htm

Jorgensen, G. (2006). Kohlberg and Gilligan: Duet or duel? *Journal of Moral Education, 35*(2), 179–196.

Kelly, M. J. (1980). *Legal ethics and legal education.* Hastings-on-Hudson: Hastings Center, Institute of Society, Ethics and the Life Sciences.

Killen, M., & Smetana, J. G. (2006). *Handbook of moral development.* Mahwah: Lawrence Erlbaum Associates.

Kohlberg, L. (1981). *Essays on moral development* (1st ed.). San Francisco: Harper & Row.

Kohlberg, L., Levine, C., & Hewer, A. (1983). *Moral stages: A current formulation and a response to critics.* Basel/New York: Karger.

Kurtines, W. M., & Gewirtz, J. L. (1984). *Morality, moral behavior, and moral development.* New York: Wiley.

Martin, M. W., & Schinzinger, R. (1996). *Ethics in engineering* (3rd ed.). New York: McGraw-Hill.

Martin, M. W., & Schinzinger, R. (2005). *Ethics in engineering* (4th ed.). Boston: McGraw-Hill.

Martin, M. W., Schinzinger, R., & Schinzinger, R. (2010). *Introduction to engineering ethics* (2nd ed.). Boston: McGraw-Hill Higher Education.

National Academy of Engineering. (2011). Online ethics center: Mission. Retrieved 3 Aug 2011, from http://www.onlineethics.org/about.aspx

National Academy of Engineering. (2013). Online ethics center for engineering and research. Retrieved 12 Mar 2013, from http://www.onlineethics.org/

National Ethics Resource Center. (2012). Ethics CORE. Retrieved 12 Mar 2013, from http://www.nationalethicsresourcecenter.net/

National Institute for Engineering Ethics. (2013). Background, mission, and purpose. Retrieved 12 Mar 2013, from http://www.niee.org/murdoughCenter/

National Institute for Engineering Ethics (Producer). (1989). *Gilbane gold.*

National Institute for Engineering Ethics (Producer). (2003). *Incident at morales.*

National Institute for Engineering Ethics (Producer). (2010). *Henry's daughters.*

National Society of Professional Engineers. (2011). NSPE code of ethics for engineers. Retrieved 4 Aug 2011, from http://www.nspe.org/Ethics/CodeofEthics/index.html

Newberry, B. (2004). The dilemma of ethics in engineering education. *Science and Engineering Ethics, 10*(2), 343–351.

Newberry, B., Austin, K., Lawson, W., Gorsuch, G., & Darwin, T. (2011). Acclimating international graduate students to professional engineering ethics. *Science and Engineering Ethics, 17*(1), 171–194.

Palmer, J., Cooper, D. E., & Bresler, L. (2001). *Fifty modern thinkers on education: From Piaget to the present.* London/New York: Routledge.

Powers, C. W., & Vogel, D. (1980). *Ethics in the education of business managers.* Hastings-on-Hudson: Hastings Center.

Rest, J. R. (1994). Background: Theory and research. In J. R. Rest & D. Narvâaez (Eds.), *Moral development in the professions: Psychology and applied ethics* (pp. 1–26). Hillsdale: L. Erlbaum Associates.

Rest, J. R. (1999). *Postconventional moral thinking: A Neo-Kohlbergian approach.* Mahwah: L. Erlbaum Associates.

Robinson, S., Dixon, R., Preece, C., & Moodley, K. (2007). *Engineering, business and professional ethics.* Amsterdam/Boston: Elsevier/Butterworth-Heinemann.

Rosen, B., & Caplan, A. L. (1980). *Ethics in the undergraduate curriculum.* New York: The Hastings Center.

Royal Academy of Engineering. (2011). Statement of ethical principles. Retrieved 11 Mar 2013, from http://www.engc.org.uk/ecukdocuments/internet/document%20library/Statement%20of%20Ethical%20Principles.pdf

Seebauer, E. G., & Barry, R. L. (2001). *Fundamentals of ethics for scientists and engineers.* New York: Oxford University Press.

Self, D. J., & Ellison, E. M. (1998). Teaching engineering ethics: Assessment of its influence on moral reasoning skills. *Journal of Engineering Education, 87*(1), 29–34.

Self, D. J., Wolinsky, F. D., Baldwin, J., & DeWitt, C. (1989). The effect of teaching medical ethics on medical students' moral reasoning. *Academic Medicine, 62*(12), 755–759.

Speight, J. G., & Foote, R. (2011). *Ethics in science and engineering*. Hoboken: Wiley.

Spodek, B., & Saracho, O. N. (2006). In B. Spodek, & O. N. Saracho (Eds.), *Handbook of research on the education of young children* (2nd ed.). Mahwah: Lawrence Erlbaum Associates.

Sutton, R. E. (1992). Review of the defining issues test. In J. J. Kramer & J. C. Conoley (Eds.), *The mental measurements yearbook*. Lincoln: Buros Institute of Mental Measurements.

Tavani, H. T. (2011). *Ethics and technology: Controversies, questions, and strategies for ethical computing* (3rd ed.). Hoboken: Wiley.

Texas A&M University. (2013). Ethics. Retrieved 12 Mar 2013, from http://ethics.tamu.edu/

Vallero, D. A. (2007). *Biomedical ethics for engineers: Ethics and decision making in biomedical and biosystem engineering*. Amsterdam/Boston: Elsevier/Academic.

van de Poel, I., & Royakkers, L. M. M. (2011). *Ethics, technology, and engineering: An introduction*. Malden: Wiley-Blackwell.

Vogt, C. M. (2008). Faculty as a critical juncture in student retention and performance in engineering programs. *Journal of Engineering Education, 97*(1), 27–36.

Weil, V. (2003). Ethics across the curriculum: Preparing engineering and science faculty to introduce ethics in their teaching. In W. Wulf (Ed.), *Emerging technologies and ethical issues in engineering: Papers from a workshop,* Oct 14–15, 2003 (pp. 79–93). Washington, DC: National Academies Press.

Yadav, A., & Barry, B. E. (2009). Using case-based instruction to increase ethical understanding in engineering: What do we know? What do we need? *International Journal of Engineering Education, 25*(1), 138–143.

Chapter 12
A Cross Cultural Comparison of Engineering Ethics Education: Chile and United States

Ruth I. Murrugarra and William A. Wallace

Abstract This chapter describes the material covered and educational approaches used in an Engineering Ethics course designed initially for an American university, and then adapted to the Chilean culture; explains the adaptation process and how cultural differences were overcome; and reports on an assessment of learning and the evolution of students' moral values using data collected throughout the course. Results from 53 students, all seniors majoring in Industrial Engineering at Universidad Adolfo Ibañez, are compared to two U.S. classes: "Ethics for Modeling for Industrial Engineering" at Rensselaer Polytechnic Institute, taught two consecutive years. The Fall 2011 class included 28 students from Industrial and Systems Engineering (ISE), 88 % seniors and 12 % juniors, and the Fall 2012 class included 26 ISE students, all seniors. After taking the class, students in both cultures were more "open to change" and the ethical theories presented were applied in case studies, both those discussed in class as well as on examinations.

Keywords Cross-cultural • Cross-national • Engineering ethics • Ethics education • Value profiles

12.1 Initial Design of the Course

The ethics course was designed as a result of a National Science Foundation (NSF) award on "Educational Simulation for Computing and Information Ethics" Robbins et al. (2009), Fleischmann et al. (2009, 2010). The objectives of this research are to assess the impact of an educational experience which aims to enhance ethical decision making and to recognize the differences in culture and values.

R.I. Murrugarra (✉)
Universidad Adolfo Ibáñez, Campus Viña del Mar, Santiago, Chile
e-mail: ruth.murrugarra@uai.cl

W.A. Wallace
Rensselaer Polytechnic Institute, Troy, NY, USA
e-mail: wallaw@rpi.edu

© Springer International Publishing Switzerland 2015
C. Murphy et al. (eds.), *Engineering Ethics for a Globalized World*, Philosophy of Engineering and Technology 22, DOI 10.1007/978-3-319-18260-5_12

This course was designed placing special emphasis on the mutually constitutive relationship between information technologies and human values, as well as the culturally situated nature of both information technologies and human values. It is novel because of the mechanism through which it is taught (case-based education through an educational simulation tool developed during the project), the range of timely and appropriate content areas (including particular emphasis on current issues such as intellectual property), and the international scope of the course (building on the notion of ethics and values as culturally constructed).

Designed initially for U.S. students, this course was taught during two consecutive years (Fall 2011 and Fall 2012) at the Department of Industrial and Systems Engineering at Rensselaer Polytechnic Institute under the name of "Ethics for Modeling for Industrial Engineering". During the Fall 2012 semester, the course was adapted to the Chilean culture and taught at the Department of Science and Engineering at Universidad Adolfo Ibañez, in Chile, under the name of "Ethics and Engineering".

12.2 Academic Design

The ethics course initially was designed for a 14-week academic semester with the goal of situating key ethical issues related to computing and information ethics in an international context. Lectures in the first half of the semester (weeks 1 to 6) are related to moral values and a variety of ethical theories; in particular, traditional Western ethics (Kantianism, social contract, utilitarianism, and virtue ethics), non-traditional Western ethics (rational egoism, ethics of care, and situated knowledge), and non-Western ethics (Classical Chinese ethics, Indian ethics, Buddhist ethics, Islamic ethics, and Ubuntu culture). In the second half of the semester (weeks 8 to 14) lectures discuss ethical issues related to information technology.

The selected course text book was *Ethics for the Information Age* by Quinn (2011). It is mostly used in the second half of the course, for the presentation and discussion of social and ethical issues related to computing technologies. But it is also used in the introductory part of the course when providing an overview of the Western ethical theories. Additional reading material is used to present the non-traditional Western ethics and non-Western ethics.

For the second half of the course, a learning tool is used to complement the material. This tool is a case-based educational computer simulation, also developed as part of the "Educational Simulation for Computing and Information Ethics" NSF project, for this course specifically and initially for U.S. students. During this part of the semester, each week a new case is discussed; its focal topic is related to the computing technology issue presented in class. Students grouped in teams use the case-based educational computer simulation to discuss and confront different ethical issues related to the Information Age using the theories learned throughout the course. A sample course calendar along with the readings used for each topic is shown in Table 12.1.

Table 12.1 Ethics course calendar

Week	Topic	Readings
1	Introduction	Bazerman and Moore 2009, Chap. 2; Banaji et al. 2003
2	Catalysts for change	Quinn 2011, Chap. 1; Willemain 1995; Little 1994
3	Values	Schwartz 2007; Friedman and Kahn 2008
4	Introduction to ethics (western ethical approaches)	Quinn 2011, Chap. 2, Sect. 8.4.2
5	Additional ethical approaches I	Bilimoria 1993; Nanji 1993; Prinsloo 1998; Smith 2006
6	Additional ethical approaches II	De Silva 1993; Hansen 1993; Haraway 2003; Held 2008
7	MIDTERM EXAM	
8	Network communications	Quinn 2011, Chap. 3
9	Intellectual property	Quinn 2011, Chap. 4
10	Privacy	Quinn 2011, Chap. 5
11	Computer and network security	Quinn 2011, Chap. 6
12	Computer reliability	Quinn 2011, Chap. 7
13	Work and wealth	Quinn 2011, Chap. 8
14	FINAL EXAM	

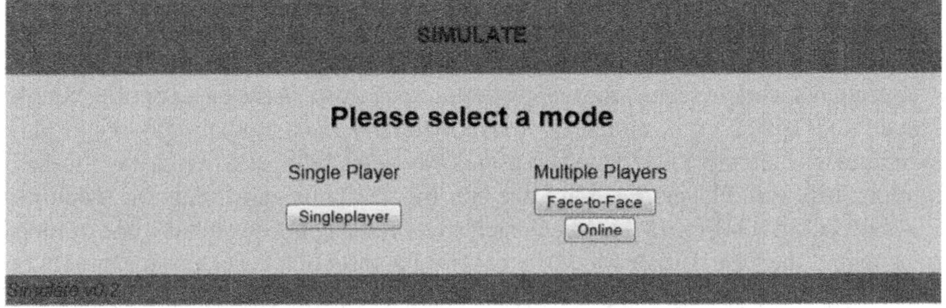

Fig. 12.1 SIMULATE SCREEN: mode selection

12.2.1   *Educational Computer Simulation Tool*

SIMULATE, the name of the case-based learning tool developed for this course, stands for $\underline{S}$imulation for [Computing] and $\underline{I}$nformation $\underline{M}$aster's [Students] to $\underline{U}$nderstand, $\underline{L}$earn, and $\underline{A}$pply $\underline{T}$eamwork and $\underline{E}$thics. It allows students to work individually or in groups with other students to resolve ethical dilemmas. It can be used in face-to-face or in online courses and exposes students to a variety of ethical principles, but also creates an environment where students can consider the relationship of values and ethical decision making (see Fig. 12.1).

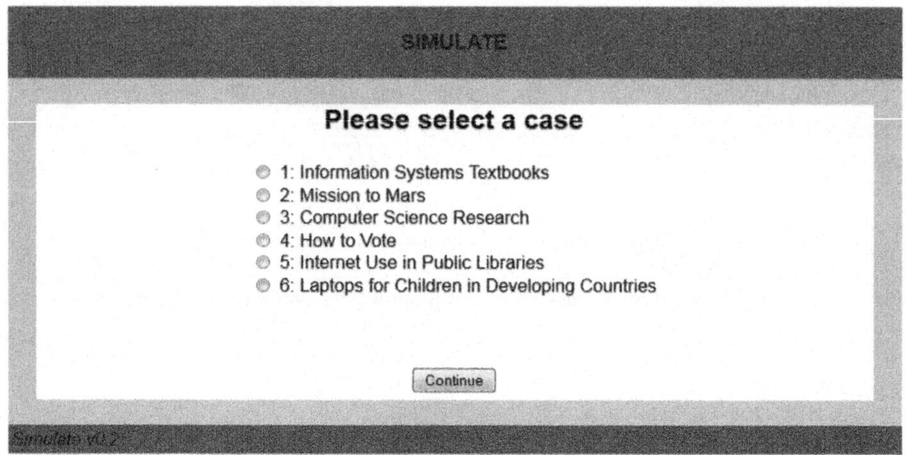

Fig. 12.2 SIMULATE SCREEN: case selection

SIMULATE presents students with a library of cases (see Fig. 12.2) where each case has a focal ethical topic. The cases cover the topical areas in the textbook (Quinn 2011): networking, intellectual property, privacy, computer and network security, computer reliability, and work and wealth.

Once a case is selected, SIMULATE presents a brief description of the roles available within a specific case. Each case consists of three different roles. Each role faces a different ethical problem within the case (see an example in Fig. 12.3).

If students work in teams, the team will play every role of the case (option "Single Player" on Fig. 12.1). If students work individually, each student chooses to play a particular role, and SIMULATE presents two possibilities: to work face-to-face (option "Multiple Players/Face-to-Face" on Fig. 12.1) or online (option "Multiple Players/Online" on Fig. 12.1), where each student then chooses to take one or more roles within the case. Either way, when solving a particular case, a role can only be played by one player, meaning either one team or one individual.

For each role on each case, the player is asked to complete a pre-decision analysis and a decision explanation, and to select the approaches used when making the decision. The pre-decision analysis is an initial impression of a role's ethical problem and what decision this particular role might take (see top-left screen on Fig. 12.4). The player is asked to discuss the different possible decisions that the role he is playing can make and enter all of the possible options as well as the implications of each possible decision.

The decision explanation is a justification of the decision the role made, based upon the options provided by the software (see top-right screen on Fig. 12.4). SIMULATE presents two ethical decisions based upon the role and the case provided. The player is asked to choose one of them and explain the factors he has considered to make that decision.

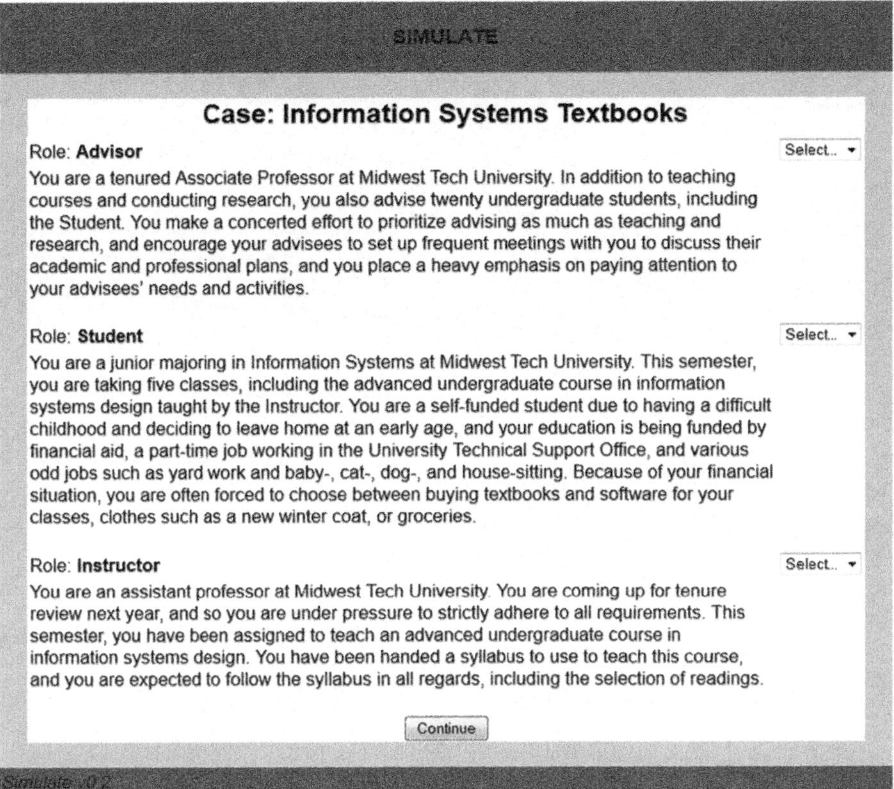

Fig. 12.3 SIMULATE SCREEN: role selection

The last screen (bottom screen on Fig. 12.4) provides a list of ethical approaches,
where the player selects the approaches used when choosing one of the decisions
provided by the software. Based on the decision chosen by the player, the software
presents the next role's ethical problem. The player then faces a sequential decision
ethical case.

12.2.2   Research\Course Evaluation Plan

In order to evaluate the educational impact and effectiveness of the course, all
students need to fill out the Schwartz Value Survey (SVS) two times during the
semester; once during the first (pre-survey) and once during the final lecture (post-
survey) of the course.

The SVS used to collect data is composed of two value lists which contain overall
56 items in terms of value descriptions. The task is to rate how important each

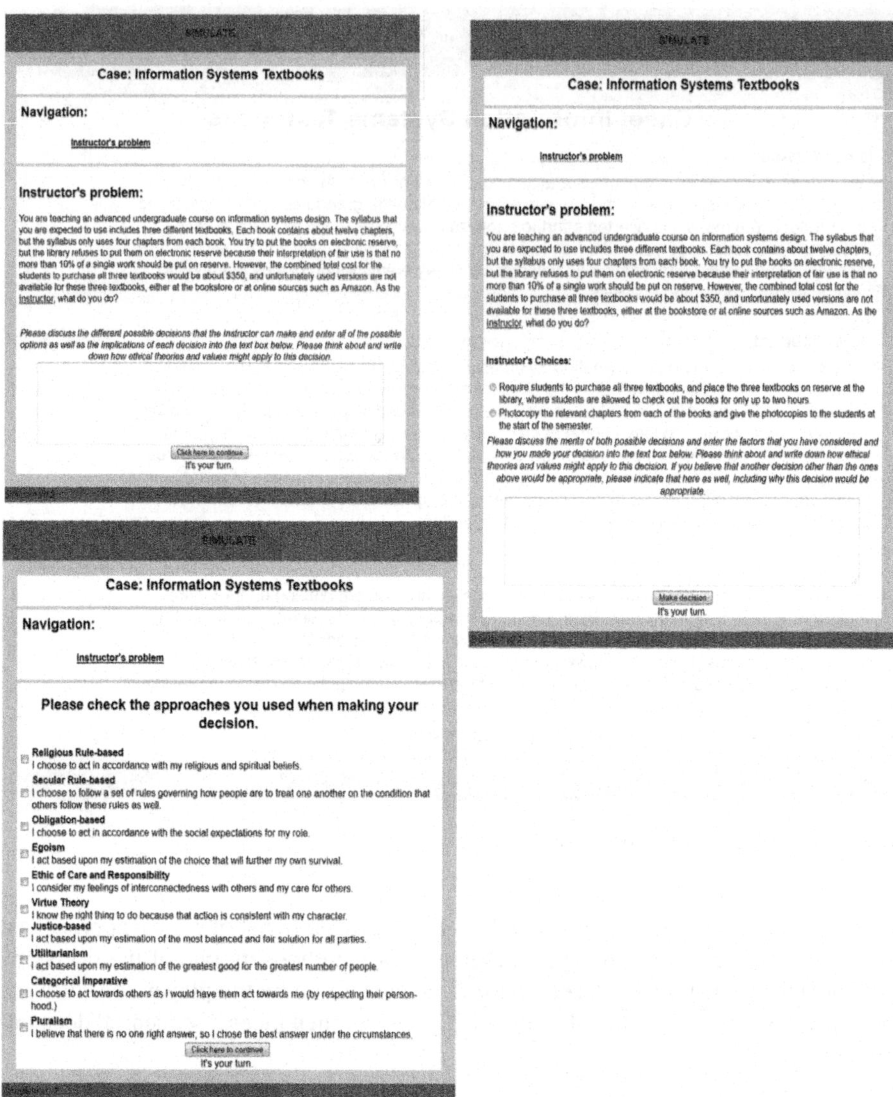

Fig. 12.4 SIMULATE SCREENS: Pre-decision, decision, and list of ethical approaches

value is for the respondent as a guiding principle in life. The importance is rated on a scale between 0 (not at all important) and 6 (very important). Additionally, outstanding values which are either opposed to the respondents' principles or of supreme importance to them can be rated with -1 or 7 respectively (Schwartz 1992, 2005).

The 56 values are grouped into ten motivational values of the individual level; the ten motivational values and their descriptions are presented in Table 12.2.

Table 12.2 Description of the ten individual motivational values

Motivational value	Description
Power	The motivational goal of power values is the attainment of social status and prestige, and the control or dominance over people and resources
Achievement	The primary goal of this type is personal success through demonstrated competence. Competence is evaluated in terms of what is valued by the system or organization in which the individual is located
Hedonism	The motivational goal of this type of value is pleasure or sensuous gratification for oneself. This value type is derived from physical needs and the pleasure associated with satisfying them
Stimulation	The motivational goal of stimulation values is excitement, novelty, and challenge in life. This value type is derived from the need for variety and stimulation in order to maintain an optimal level of activation. Thrill seeking can be the result of strong stimulation needs
Self-direction	The motivational goal of this value type is independent thought and action (for example, choosing, creating, exploring). Self-direction comes from the need for control and mastery along with the need for autonomy and independence
Universalism	The motivational goal of universalism is the understanding, appreciation, tolerance, and protection of the welfare of all people and of nature
Benevolence	The motivational goal of benevolent values is to preserve and enhance the welfare of people with whom one is in frequent personal contact. This is a concern for the welfare of others that is more narrowly defined than universalism
Tradition	The motivational goal of traditional values is respect, commitment, and acceptance of the customs and ideas that one's culture or religion imposes on the individual. A traditional mode of behavior becomes a symbol of the group's solidarity and an expression of its unique worth and, hopefully, its survival
Conformity	The motivational goal of this type is restraint of action, inclinations, and impulses likely to upset or harm others and violate social expectations or norms. It is derived from the requirement that individuals inhibit inclinations that might be socially disruptive in order for personal interaction and group functioning to run smoothly
Security	The motivational goal of this type is safety, harmony, and stability of society or relationships, and of self

These values have different underlying motivations. Figure 12.5 shows the four dimensions or underlying motivations that Schwartz (1992, 2005) found. The closer the motivational values, the more similar are their underlying motivations. The further apart the motivational values, the more antagonistic are their underlying motivational values.

In addition, text data from SIMULATE, in particular, the pre-decision and decision explanations, the ethical decision made, and the ethical approaches selected for each role and case from every player were recorded. These two sets of data, text data from SIMULATE and pre- and post-surveys are used to measure the impact of SIMULATE and the ethics course on learning and attitudes.

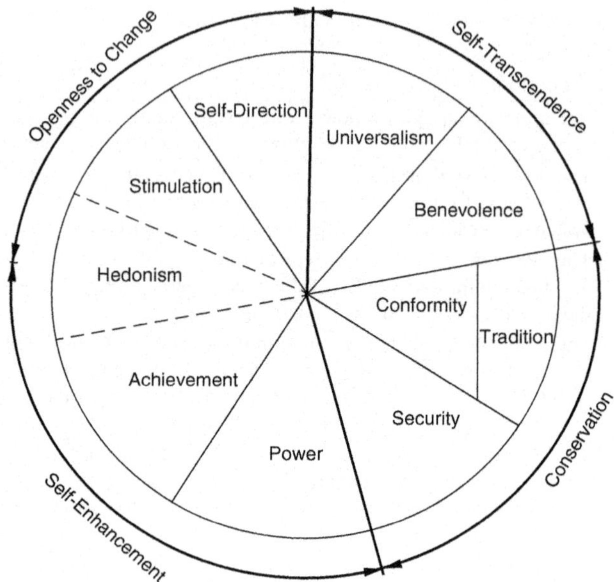

Fig. 12.5 Dimensions or underlying motivations of the ten motivational values

It is important to note that changes on value profiles could be impacted by externalities, such as social interaction, or media influence. This study does not take into consideration these externalities. Future work will test changes on value profiles on groups of Industrial Engineering students that have not taken the ethics class.

12.3 Chilean Ethics Course Adaption Process

This educational research was reproduced in an ethics course at the Adolfo Ibañez University in Chile, a Spanish-speaking country in South America, where the course is elective and taught in Spanish to senior students with majors in Industrial Engineering and Engineering Management. Adolfo Ibañez University is the ninth best university in Chile according to the overall Chilean ranking. It is also placed fifth among the ten best universities in Industrial Engineering.

12.3.1 Overview

Chile has 180 institutions that provide higher education; 59 of them are colleges or universities. The first 30 universities according to the Chilean ranking, which include the first ten universities in Industrial Engineering, held almost 60 % of the students seeking higher education.

Among those 30 universities, only 36 % of them have a mandatory course concerning ethics. "Professional Ethics" is the ethics course most often taught, and it is usually taught from a Christian point of view.

The most frequent topics included in the current Chilean ethics programs are: ethics and society, morals and ethics, ethics principles, occidental ethical theories such as utilitarianism, Kantianism, virtue ethics (Aristotle), and the code of ethics of an engineer.

12.3.2 Textbook and Complementary Readings

The Spanish ethics course does not use a textbook; readings from several books and journals are used to complement the material shown during lectures. When providing an overview of the Western ethical theories, such as, subjectivism (Rachels 1993), relativism (Wong 1993), Kantianism (O'neill 1993), utilitarianism (Goodin 1993), and social contract (Kymlicka 1993), we used the Spanish version of the book *A Companion to Ethics* by Singer et al. (1995).

The topics studied under "Additional Ethical Approaches" include: Classical Chinese ethics, Indian ethics, Buddhist ethics, Islamic ethics, Ubuntu culture, rational egoism, feminism, and gender identity.

To provide an overview of non-Western ethics, students read essays written by different authors gathered in the Spanish version of *A Companion to Ethics* (1995), the titles of which can be translated as "Classical Chinese ethics" (Hansen 1993), "Indian ethics" (Bilimoria 1993), "Buddhist ethics" (De Silva 1993), and "Islamic ethics" (Nanji 1993).

For the topic of Ubuntu culture, the essay "Ubuntu culture and participatory management" (Prinsloo 1998) was replaced by a Spanish essay, the title of which can be translated as "Ubuntu as a model of restorative justice; an African contribution to the equality and human dignity's debate" (Kakozi 2011).

Regarding the topic of rational egoism, the essay "The Egoism" (Baier 1993) from the Spanish version of *A Companion to Ethics* was used along with a Spanish essay, the title of which can be translated as "Ayn Rand and the moral principles of Liberalism" (Etxebarria 2004). The first reading introduces the reader to the different types of egoism (psychological egoism, ethical egoism, rational egoism) and the second reading focuses on the moral principles of Ayn Rand's philosophy, in which rational egoism is perceived as a virtue.

For the topic of feminism, we used the Spanish version of the reading material of Haraway (1995) found in the Spanish version of the book *Simians, Cyborgs, and Women: The Reinvention of Nature*.

And, for the last additional ethical approach related to gender identity, the essay "Gender identity and the ethics of care in globalized society" (Held 2008) has been replaced by two readings, the titles of which can be translated as: "Globalization and new cultural identities: Gender identity under construction" (Perez 2011) and "Ethics of care and ethics of justice under Carol Gilligan's moral theory" (Fascioli 2010).

The social and ethical issues related to computing technologies, which are presented and discussed during the second half of the course, are contemporary topics which the students are exposed to in their daily life. We do not use a textbook for this part; instead we have discussions in class and presentations of real ethical cases related to the subject.

12.3.3 Case-Based Ethical Problems

As mentioned before, a small percentage of Chilean universities have a mandatory course concerning ethics. And the universities that have an existing ethics course in their program usually follow a philosophical or theological perspective from a Christian point of view, not a case-based approach. Hence, it is not surprising that very few ethical cases with Chilean context have been developed; and that the few courses that use cases rely on American ones.

SIMULATE was used to present ethical cases to the students and facilitate their discussion. The educational simulator contains six cases:

1. Information Systems Textbooks
2. Mission to Mars
3. Computer Science Research
4. How to Vote
5. Internet Use in Public Libraries
6. Laptops for Children in Developing Countries

The cases were presented in English; no translation was made since all students have a medium-to-advanced level of English as required by the Adolfo Ibañez University. But, five of the six cases needed further explanation because of terms only used or known in the U.S. The first five cases are situated in an American context, such as academia in the U.S. or the U.S. government. Only one case, the sixth, was situated in a developing country, more specifically in a Sub-Saharan African nation. This was the only case where it wasn't necessary to have additional explanatory information. Details about which situations and/or terms needed to be explained in each of the five ethical cases are elaborated in this section.

Case 1, "Information Systems Textbooks", involves three roles: the advisor, who is a tenured associate professor; the instructor, who is an assistant professor coming up for tenure review; and the student, who is a junior at a certain university. The concept of tenure was new to the students; moreover, they were not familiar with U.S. faculty hierarchy terms. Also, they were not familiar with the meaning of the U.S. students' classification, and could not relate those concepts to their own experiences, since most university degrees are 6 years long in Chile compared to 4 years in the U.S.

Furthermore, in one of the scenarios, the instructor's problem is that he is teaching an advanced undergraduate course and required to use chapters from three different textbooks. The books are expensive, and unfortunately used versions are not available for these textbooks, either at the bookstore or online. He cannot put the books on electronic reserve, due to the library's interpretation of fair use. The term "electronic reserve" was new to the students and needed to be explained in order for them to understand the options of the instructor.

Case 2, "Mission to Mars", has one of the roles played by the division director, who has a B.S. in computer science from a well-known institution and a Ph.D. in software engineering from one of the top U.S. universities. The majority of students were not familiar with the academic abbreviations and their exact meanings. Also, not all were familiar with the academic institutions mentioned in the case, and they did not know why or for which academic accomplishments they are well-known. This prior knowledge enriches the context of the case, and was missing in most of the Chilean students. In addition, the following terms were not known or not remembered at the time of solving the case and had to be explained, but this was not exclusive of Chilean students: decision support systems, expert systems, and primary data.

Case 3, "Computer Science Research", has three roles: FBI director, who needs to balance the need for safety through counterterrorism efforts with the rights to privacy guaranteed by the U.S. Bill of Rights; a student, who is pursuing a Ph.D. and is a strong advocate of personal privacy; and a professor, who needs to balance accommodating his students' moral preferences while ensuring the successful completion of his research projects. The students needed further explanation about the Bill of Rights, since it is part of the U.S. Constitution. Also, they required an explanation of the meaning of different academic abbreviations and the U.S. faculty hierarchy terms.

Case 4, "How to Vote", in general, needed an initial brief explanation of the voting system in the U.S., so students could be situated in the context of the ethical dilemma.

Finally, Case 5, "Internet Use in Public Libraries", involves the following roles: the library director, who is in charge of a small public library and is a member of the American Library Society (ALA); the librarian, who works at the reference desk at this small public library and needs to consider the ALA Code of Ethics, as well as the best interests of the library, his fellow staff members, and all of his patrons when fulfilling his responsibilities; and a U.S. representative, representing a conservative rural district in a Southeastern state. The students needed additional information about the ALA Code of Ethics and its principles. In addition, within the case, the U.S. House of Representatives is mentioned, thus, the students required a brief elaboration to completely understand the situation.

12.3.4   Schwartz Value Survey

According to the Schwartz Value Survey user manual (Schwartz 2009), "you will need to know the first language of the participants and administer the survey in that language..." For that reason, a validated Spanish version of the Schwartz Value Survey (Clemente Díaz, Gouveia, & Vidal 1998) was used to assess moral values.

12.4 Data and Text Analysis

12.4.1 Schwartz Data Analysis

According to the Schwartz Value Survey user manual (Schwartz 2009), surveys with at least one of the following characteristics must be excluded from the analysis:

- 15 or more value items blank
- more than 30 % blanks from the value items related to a particular motivational value
- used a particular scale anchor 35 times or more

None of the surveys taken in both the U.S. and Chilean ethics classes were removed due to these characteristics.

During the study, additional inconsistencies were found on particular value items:

- scores greater than 7 or smaller than -1
- non-numeric characters as scores
- value items scored twice
- value items not scored

The first two inconsistencies indicate that the student did not follow the guidelines of the survey; hence surveys having any of the first two items were dropped from the analysis. Only 3 % of the total surveys were not used due to these characteristics. On the other hand, surveys containing the last two inconsistencies were kept for analysis, where only the inconsistent value item was removed. There were no value items that were largely unanswered or scored two times.

In addition, surveys from subjects who did not complete the survey both times (pre- and post-surveys) were kept for the analysis. In cross-cultural samples, each individual is not necessarily independent from the rest of the population; hence deleting the case could introduce bias to our estimations and errors (Dow and Eff 2009).

Each individual uses the response scale differently, which can provide misleading results. To account for these differences, a scale correction called the MRAT (mean rating of a particular individual) was applied (Schwartz 2009). The MRAT is computed taking the average of 56 value items for each individual. Then, the difference between the score of a value item and the MRAT is computed. These differences are used for the analysis.

To compute the individual score associated with each motivational value, a simple average of the value items associated with each motivational value were performed. The list of value items per motivational value is shows in Table 12.3.

Table 12.3 Associations
between value items and
motivational values

Motivational value	SVS value items
Power	3,12,23,27,46
Achievement	14,34,39,43,48,55
Hedonism	4,50
Stimulation	9,25,37
Self-direction	5,16,31,41,53
Universalism	1,2,17,24,26,29,30,38
Benevolence	6,10,19,28,33,45,49,54
Tradition	18,32,36,44,51
Conformity	11,20,40,47
Security	7,8,13,15,22,42,56

12.4.2 Nationality and Ethnicity Influence

To identify cross-national and/or cross-ethnic differences in value profiles, students were categorized into groups according to their citizenship and country of birth (Kankaras and Moors 2012). In the U.S., students were divided into three groups: American citizens, immigrant citizens, and immigrant non-citizens. Due to the small sample size of the last group, this research does not make a distinction among immigrants coming from different cultures or countries. In the Chilean class, all students were Chilean citizens.

Two types of comparisons were performed: a comparison of value profiles among groups and classes, one for the pre-survey and one for the post-survey; and a comparison between pre- and post-survey for each group and class.

In order to analyze differences in personal values across groups (first comparison), the Kruskal-Wallis test is used. To measure the changes in moral values from the beginning to the end of the course (second comparison), the Wilcoxon signed-ranked test is used.

The Kruskal-Wallis test is a non-parametric test that compares three or more independent groups, and does not need to assume that the sources come from a normal distribution. It is robust for groups with different sample sizes, and can work with ordinal data (rating-scale data). The null hypothesis for this test is that "there is no significant difference among the medians of the k-groups". A small p-value rejects the null hypothesis, which means that at least one group's median differs from one of the others.

To further identify which group is different, a Dunn's post-test is performed. This test compares the difference in the median for each group pair. If the statistic of the pair is greater than the critical value, then the null hypothesis that "there is no significant difference between the medians of the pair" is rejected. With smaller sample sizes the power of the Kruskal-Wallis test decreases, so these results need to be analyzed carefully.

The Wilcoxon signed-ranked text is also a non-parametric test that compares two paired samples. It does not assume that the data comes from a normally distributed population, works with ordinal data (rating-scale), and is used for "before" and "after" data. This test compares the median from two dependent samples (pre-survey and post-survey); if the p-value is small, the null hypothesis that the medians do not differ significantly is rejected. Since it is a rank-based test, it is robust to outliers. The minimum sample size for which the statistical W can be computed is five (Lowry 1998–2011).

12.4.3 Team Interaction Influence

To identify cross-team differences in value profiles, students were categorized into groups according to the four-member teams that they formed for the analysis of ethical cases using SIMULATE. We compute three team scores associated with each motivational value; the mean, the median, and the variance; which were computed as the simple average, the median, and the sample variance of the individual motivational values of the team members respectively.

Two types of comparisons were performed: a comparison of mean value profiles, median value profiles, and variance of the value profiles among teams and classes, one for the pre-survey and one for the post-survey; and a comparison between pre- and post-survey for each team and class.

The tests performed were the same used for the analysis of individual motivational values. The Kruskal-Wallis test and the Dunn's post-test are used for the first comparison, and the Wilcoxon signed-ranked test is used for the second comparison.

12.5 SIMULATE Exploratory Text Analysis

12.5.1 Consistency Among Answers

Text data from SIMULATE comes from the analysis of ethical cases in teams of four members. Each case consists of three scenarios; where the content of the current scenario depends on the decision made on the previous scenario. For each scenario, a team records a pre-decision, a decision, and the ethical approaches used in the analysis. In the pre-decision and the decision, the team has to think and write down ethical theories and values that might apply to the ethical decision. Finally, the team has to select the approaches they used when making their decision. Different ethical theories offer different approaches. The list of approaches to ethical theories taught in this course is shown in Table 12.4.

Consistency among these three sources of text information is explored to test for coherent and reliable answers. To measure consistency, the name of the ethical theories written by the teams in their pre-decision and decision analysis were extracted and compared to the ethical approaches selected from the list. Based on this information, this research found five different groups of answers.

Table 12.4 Approaches to ethical theories

Ethical approach	Ethical theories
Utilitarianism	Act utilitarianism
	Rule utilitarianism
Egoism	Rational egoism
Ethics of care and responsibility	Ethics of care
Secular rule-based	Social contract
Obligation-based	Social contract
Categorical imperative	Kantianism
Virtue theory	Confucianism
	Buddhism
Pluralism	Act utilitarianism
Justice-based	Social contract
Religious rule-based	Islamism
	Hinduism
	Divine command

The first group explicitly named an ethical theory either in their pre-decision or in their decision but not in both, and did not select the ethical approach related to that particular ethical theory. The second group selected an ethical approach but did not name it in the pre-decision nor the decision. The third group named the ethical theory in both pre-decision and decision, but did not select the ethical approach related to that particular ethical theory. The fourth group explicitly named an ethical theory either in their pre-decision or in their decision but not in both, and did select the ethical approach related to that particular ethical theory. The fifth and last group named an ethical theory in both their pre-decision and decision, and did select the ethical approach related to that particular ethical theory. Groups where a particular ethical theory was mentioned in two of the three sources of text information were called "consistent".

12.5.2   Influence by Team and by Case

In statistics, the two-way analysis of variance (ANOVA) test examines the influence of different categorical independent variables on one dependent variable. The two-way ANOVA can determine the main effect of contributions of each independent variable and also identify if there is a significant interaction effect between the independent variables. The preliminary text study examines if the consistency of the answers is influenced by team and/or by case, and if the ethical theory chosen to analyze an ethical dilemma is influenced by team or by case. The influence of gender, year of study (senior, junior, etc.), country of birth, and value profile of the team members were not examined due to a small sample size.

12.5.3 Comparison of Commonly Used Terms

Different ethical theories are based on different principles, approaches, and ideas. Since they are different, an explanation of the ethical theory or an argument involving a particular ethical theory may involve particular words, phrases, or terms associated with it.

This research finds the most used words and phrases associated with each ethical theory, as a way to measure the learning process. In text mining, this is called the bag of words (BOW) approach (Feldman and Sanger 2007), where the system represents each document as a weighted vector of terms. The weight associated with each term is the number of times that a term appears in the document.

The pre-processing of the data consisted of the correction of spelling, the exclusion of small, common words such as "it" and "the", and the grouping of words by root. Once the pre-processing stage was completed, for each ethical theory, all pre-decision and decision answers where that particular ethical theory was mentioned were extracted. The most frequent words on each extracted text were found.

12.6 Results

12.6.1 Schwartz Data Analysis Results

First, we compared individual values among American citizens, immigrant citizens/residents, and Chilean citizens, to verify the hypothesis that students with different nationalities and ethnicities have different value profiles.

We found that American citizens and immigrant citizens/residents value profiles were very different. American citizens highly rate values related to personal interest like Achievement and Hedonism, while immigrant citizens/residents highly rate values related to collective interest like Tradition (see Fig. 12.6). It is important to note that most of the students in the immigrant citizens/residents group were from Asian countries.

On the other hand, Chilean citizens have a value profile more similar to American citizens than to the immigrant citizens/residents group. Chilean citizens, like U.S. citizens, also highly rate values related to personal interest; the difference is that U.S. citizens value more Achievement while Chilean citizens value more Hedonism. On the collective interest side, both groups give lower rates, but Chilean students value more Universalism while U.S. citizens value more Conformity (see Fig. 12.7).

To measure the impact of the ethics class in the value profiles of the different groups, we compared individual value profiles obtained before and after the course. We found that there are different statistically significant changes depending on the nationality and ethnicity of the students.

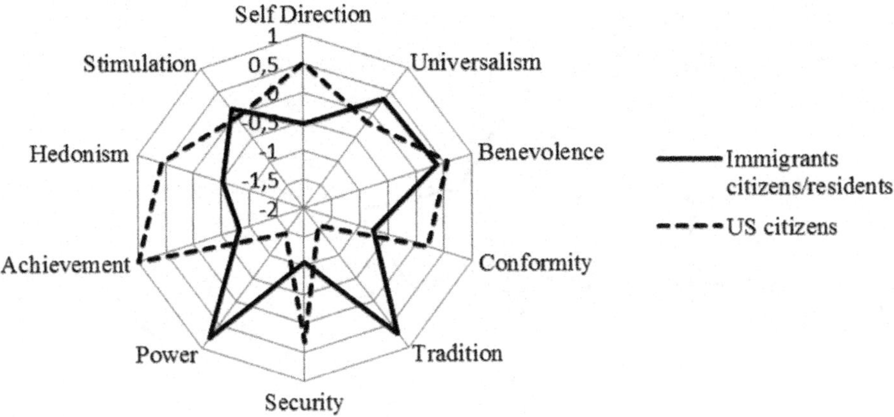

Fig. 12.6 Initial differences between U.S. citizens and immigrant citizens/residents

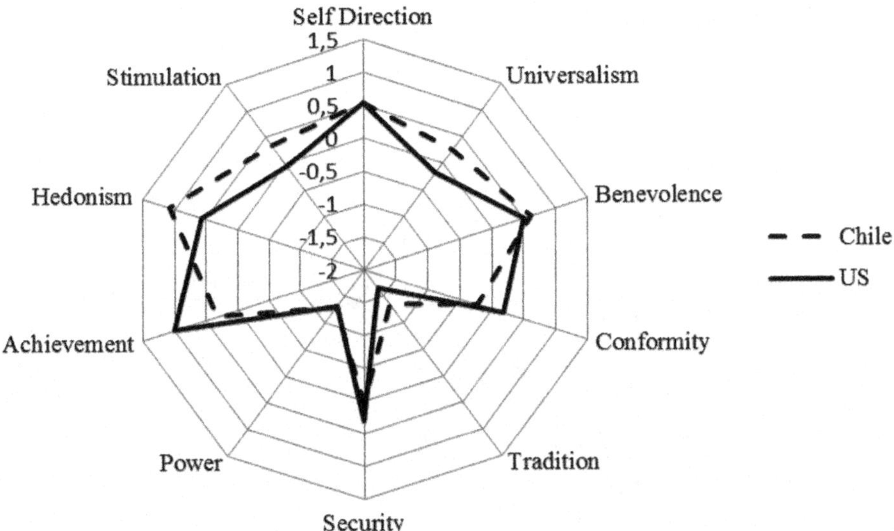

Fig. 12.7 Initial differences between U.S. citizens and Chilean citizens

For the ethics class taught in the U.S., for American citizens, the results show that after taking the ethics class, this group values Achievement less; which is a value that represents personal success, but also values Conformity less; which is a value related to collective interest. There is no statistically significant change on the other values related to collective interest (Benevolence, Tradition) and neither on the values that represent mixed interest values (Universalism and Security) (see Fig. 12.8).

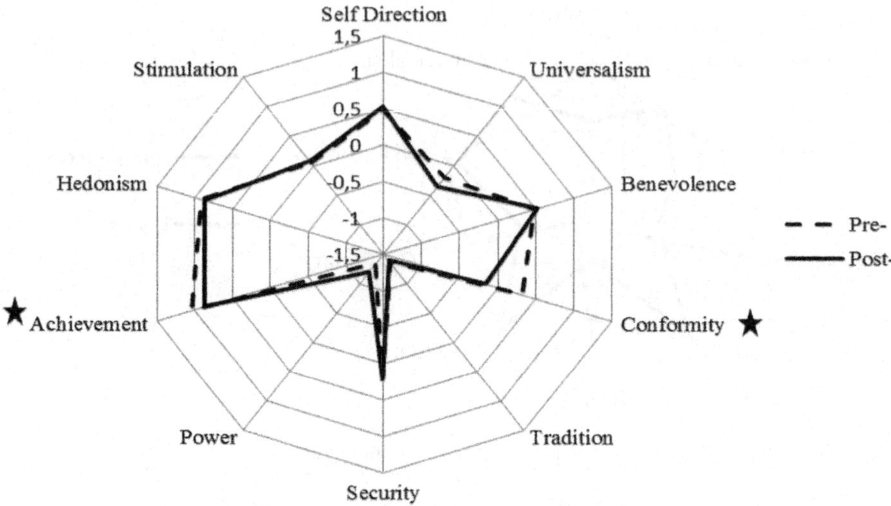

Fig. 12.8 Differences in U.S. citizens values between pre- and post-survey

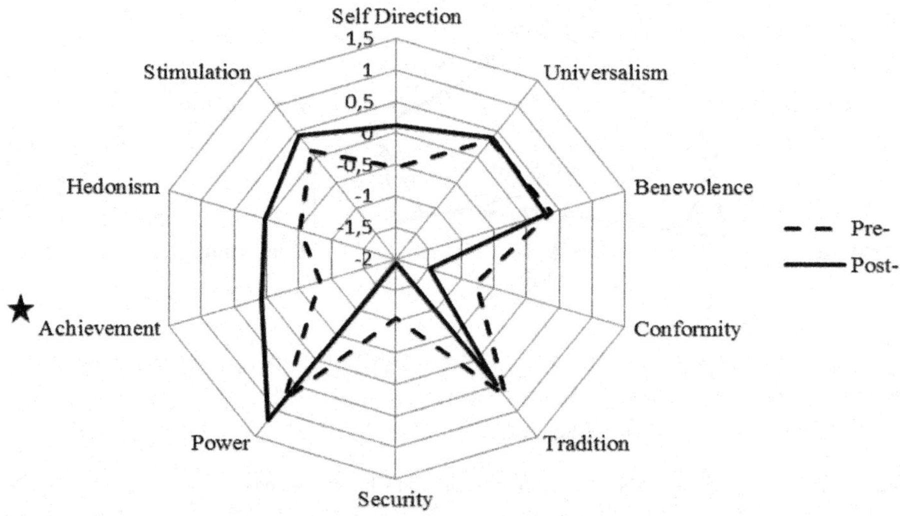

Fig. 12.9 Differences in U.S. residents values between pre- and post-survey

On the other hand, immigrant citizens/residents value more motivational values related to personal interests like Achievement after taking the ethics class. There were no statistically significant changes in the collective side (see Fig. 12.9).

For the ethics class taught in Chile, where all students were Chilean citizens, the results show that after taking the class, they value less Stimulation (value related to personal interests) and value more Benevolence (value related to collective interest) (see Fig. 12.10).

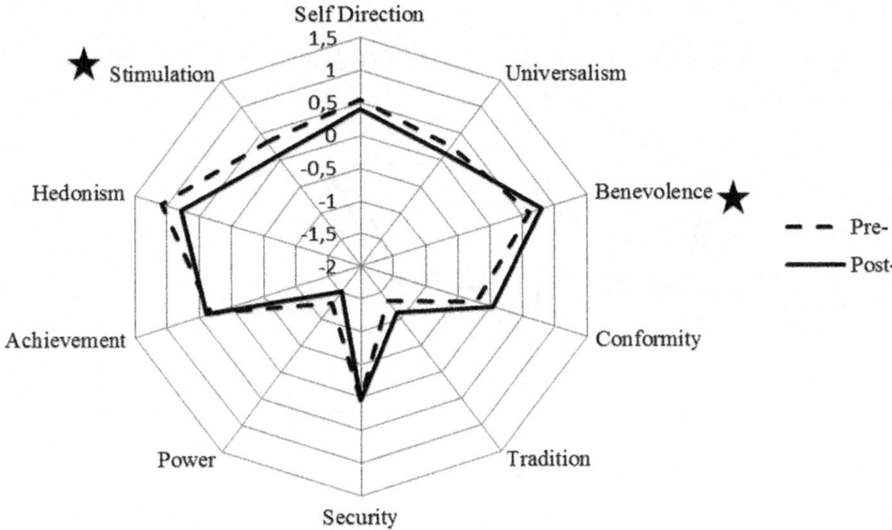

Fig. 12.10 Differences in Chilean citizens values between pre- and post-survey

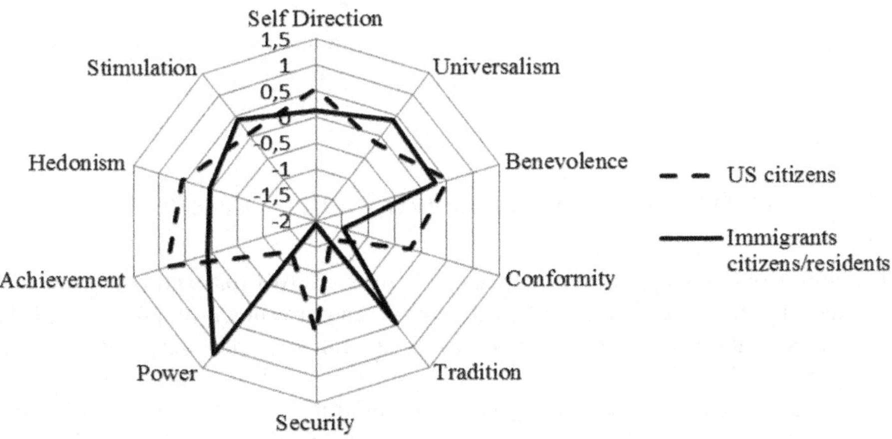

Fig. 12.11 Final differences between U.S. citizens and immigrant citizens/residents

After taking the ethics class, the results show that American citizens and immigrant citizens/residents' value profiles were not as different as prior to taking the class. Immigrant citizens/residents rank higher values related to Achievement, approaching the scores obtained from American citizens. However, Tradition was still a significant difference (see Fig. 12.11).

When analyzing the final differences between U.S. citizens and Chilean citizens, there were still differences in both personal-related values and collective-related values (see Fig. 12.12).

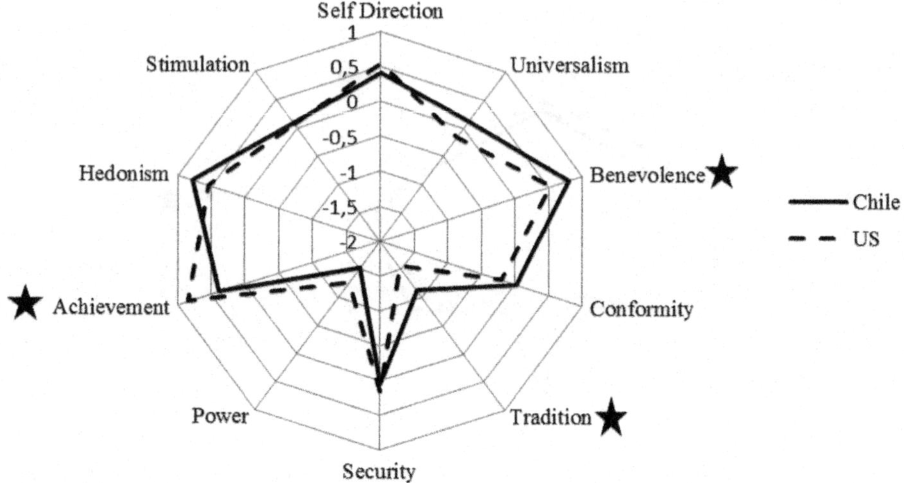

Fig. 12.12 Final differences between U.S. citizens and Chilean citizens

Table 12.5 BOW of most used ethical theories

Ethical theories	Most used words
Social contract	Fair, laws, social
Utilitarianism	Benefit, outweigh
Kantianism	Ought
Rational egoism	Best, ego

12.6.2   SIMULATE Exploratory Text Analysis Results

The most commonly used terms or bag of words (BOW) for each ethical theory are the same for both U.S. and Chilean classes, and are shown in Table 12.5. It is interesting that the words that represent an ethical theory are related with its core principles.

Social contract, for example, focuses on authority in society and agreements in the society; and its BOW is related with society ("social") and its agreements ("laws") and the idea of justice or equality in the society ("fair"). Utilitarianism focuses on the concept of utility, comparing benefits or happiness and suffering. Its BOW refers to the existence of benefits ("benefit") and that depending on whether the utility is positive or negative, a decision could be ethically correct or incorrect ("outweigh"). Kantianism focuses on ethical duties based on two different principles. The idea of duty is reflected in the bag or words ("ought"). Rational egoism focuses on the maximization of our own happiness. The BOW reflects the idea of individualism ("ego"), and the desire to do what is best for the individual ("best"). All other ethical theories did not have enough data to create a BOW for each of them.

For both classes, certain cases influence the use of social contract when analyzing the decision: Case 4 "How to Vote" and Case 5 "Internet Use in Public Libraries". It is important to note that both cases discuss social issues. Cases do not influence the utilization of all remaining ethical theories when solving the ethical dilemmas. This study also finds that teams do not have significant influence on the utilization of any ethical theory when discussing and analyzing the ethical problems and when arriving at a decision.

12.7   Conclusions

This work provides an explanation of the adaptation process of an Engineering Ethics course designed initially for an American university to the Chilean culture and how cultural differences were overcome; and a preliminary analysis about the effects of an ethics course on student moral values. Although more data is necessary to have stronger conclusions, we can point out some interesting insights:

- In order to duplicate the educational experience in a non-English speaking culture, the textual material could be the same, the only difference being the language.
- On any ethics cases used in the courses that is not culture neutral, culture-specific information must be explained.
- There were differences in motivational values among groups of different nationality and ethnicity, as measured by the Schwartz Value Survey taken at the beginning of the semester.
- There were changes in motivational values as measured by the Schwartz Value Survey from the beginning of the course until the end. However, the motivational values that changed were different between U.S. and non-U.S. classes, probably due to cultural differences.
- A particular ethical theory has specific words, phrases, or terms associated with it, independent of the language of the argument that uses it.
- Cases with social context and content influence the utilization of Social Contract theory when analyzing the decision.

Acknowledgements This material is based upon work supported by the National Science Foundation under Grant 0734894. The authors would also like to thank Jonathan Taylor for his valuable work on the collection, formatting, analysis, and revision of the data.

References

Baier, K. (1993). Egoism. In P. Singer (Ed.), *A companion to ethics* (pp. 281–290). Oxford: Blackwell.

Banaji, M., Bazerman, M., & Chugh, D. (2003). How (un)ethical are you? *Harvard Business Review, 81*(12), 56–64.

Bazerman, M., & Moore, D. (2009). *Judgment in managerial decision making* (7th ed.). Hoboken: Wiley.

Bilimoria, P. (1993). Indian ethics. In P. Singer (Ed.), *A companion to ethics* (pp. 43–57). Oxford: Blackwell.

Clemente Díaz, M., Gouveia, V., & Vidal, M. (1998). El cuestionario de valores de Schwartz (CVS) propuesta de adaptación en el formato de respuesta. *Revista de psicología social, 13*(3), 463–469.

De Silva, P. (1993). Buddhist ethics. In P. Singer (Ed.), *A companion to ethics* (pp. 58–68). Oxford: Blackwell.

Dow, M., & Eff, A. (2009). Multiple imputation of missing data in cross-cultural samples. *Cross-Cultural Research, 43*(3), 206–229.

Etxebarria, G. (2004). Ayn Rand y los fundamentos morales del liberalismo [Ayn Rand and the moral principles of Liberalism]. *Cuadernos de pensamiento político, 4*, 187–198.

Fascioli, A. (2010). Ética del cuidado y ética de la justicia en la teoría moral de Carol Gilligan [Ethics of care and ethics of justice under Carol Gilligan's moral theory]. *Revista del Departamento de Filosofía de la Práctica, 12*, 41–57.

Feldman, R., & Sanger, J. (2007). *The text mining handbook: Advanced approaches in analyzing unstructured data.* Cambridge/New York: Cambridge University Press.

Fleischmann, K. R., & Wallace, W. A. (2010). Value conflicts in computational modeling. *Computer, 43*(7), 57–63.

Fleischmann, K. R., Robbins, R. W., & Wallace, W. A. (2009). Designing educational cases for intercultural information ethics: The importance of diversity, perspectives, values, and pluralism. *Journal of Education for Library and Information Science, 50*(1), 4–14.

Friedman, B., & Kahn, P. H., Jr. (2008). Human values, ethics, and design. In J. A. Jacko & A. Sears (Eds.), *The human-computer interaction handbook* (2nd ed., pp. 1241–1266). New Jersey: Lawrence Erlbaum Associates.

Goodin, R. (1993). Utility and the good. In P. Singer (Ed.), *A companion to ethics* (pp. 337–346). Oxford: Blackwell.

Hansen, C. (1993). Classical Chinese ethics. In P. Singer (Ed.), *A companion to ethics* (pp. 69–81). Malden: Blackwell.

Haraway, D. (1995). *Conocimientos situados: La cuestión científica en el feminism y el privilegio de la perspectiva parcial* [Situated knowledges: The science question in feminism and the privilege of partial perspective] (Ciencia, cyborgs y mujeres: La reinvención de la naturaleza, pp. 313–425). Madrid: Ediciones Cátedra.

Haraway, D. (2003). Situated knowledges: The science question in feminismo and the privilege of partial perspective. In C. R. McCann & S.-K. Kim (Eds.), *The feminist theory reader: Local and global perspectives* (pp. 391–403). New York: Routledge.

Held, V. (2008). Gender identity and the ethics of care in globalized society. In R. Whisnant & P. DesAutels (Eds.), *Global feminist ethics* (pp. 43–57). Lanham: Rowman & Littlefield Publishers.

Kakozi, J. (2011). *Ubuntu como modelo de justicia restaurativa. Un debate africano al debate sobre la igualdad y la dignidad humana* [Ubuntu as a model of restorative justice; an African contribution to the equality and human dignity's debate]. XII International Congress of the Latin American Association of African and Asian studies. Bogota, Colombia.

Kankaras, M., & Moors, G. (2012). Cross-national and cross-ethnic differences in attitudes: A case of Luxembourg. *Cross-Cultural Research, 46*(3), 224–254.

Kymlicka, W. (1993). The social contract tradition. In P. Singer (Ed.), *A companion to ethics* (pp. 267–280). Oxford: Blackwell.

Little, J. (1994). On model building. In W. A. Wallace (Ed.), *Ethics in modeling* (pp. 167–182). Oxford: Pergamon.

Lowry, R. (1998–2011). Concepts and applications of inferential statistics. http://vassarstats.net/textbook. Accessed 2 Nov 2013.

Nanji, A. (1993). Islamic ethics. In P. Singer (Ed.), *A companion to ethics* (pp. 106–118). Oxford: Blackwell.

O'neill, O. (1993). Kantian ethics. In P. Singer (Ed.), *A companion to ethics* (pp. 253–266). Oxford: Blackwell.

Pérez, F. (2011). Globalización y nuevas identidades culturales: Identidad de género en construcción [Globalization and new cultural identities: Gender identity under constructions]. *Revista Sociedad y Equidad, 2*, 66–85.

Prinsloo, E. D. (1998). Ubuntu culture and participatory management. In P. H. Coetzee & A. P. J. Roux (Eds.), *The African philosophy reader* (pp. 41–51). London/New York: Routledge.

Quinn, M. J. (2011). *Ethics for the information age* (4th ed.). Boston: Addison Wesley.

Rachels, J. (1993). Subjectivism. In P. Singer (Ed.), *A companion to ethics* (pp. 581–592). Oxford: Blackwell.

Robbins, R. W., Fleischmann, K. R., & Wallace, W. A. (2009). Computing and information ethics: Challenges, education, and research. In R. Luppicini & R. Adell (Eds.), *Handbook of research on technoethics* (pp. 391–408). New York: The Idea Group.

Schwartz, S. H. (1992). Universals in the content and structure of values: Theory and empirical tests in 20 countries. In M. Zanna (Ed.), *Advances in experimental social psychology* (Vol. 25, pp. 1–65). San Diego, California: Academic.

Schwartz, S. H. (2005). Basic human values: Their content and structure across countries. In A. Tamayo & J. B. Porto (Eds.), *Valores e comportamento nas organizações* [Values and behavior in organizations] (pp. 21–55). Brazil: Vozes.

Schwartz, S. H. (2007). Value orientations: Measurement, antecedents, and consequences across nations. In R. Jowell, C. Roberts, R. Fitzgerald, & G. Eva (Eds.), *Measuring attitudes cross-nationally: Lessons from the European Social Survey* (pp. 169–203). London: Sage.

Schwartz, S. H. (2009). Draft user's manual: Proper use of the Schwartz value survey, version 14 Jan 2009, compiled by Romie Littrell. Auckland, New Zealand: Centre for Cross Cultural Comparisons, http://www.crossculturalcentre.homestead.com. Accessed 2 Nov 2013.

Singer, P., Rubio, J. V., & Vigil, M. (1995). *Compendio de ética* [A companion to ethics]. Madrid: Alianza Editorial.

Smith, T. (2006). Rational egoism: A profile of its foundations and basic character. In *Ayn Rand's Normative ethics* (pp. 19–47). Cambridge, UK: Cambridge University Press.

Willemain, T. (1995). Model formulation: What experts think about and when. *Operations Research, 43*(6), 916–932.

Wong, D. (1993). Relativism. In P. Singer (Ed.), *A companion to ethics* (pp. 593–604). Oxford: Blackwell.

Chapter 13
Responsible Conduct of Research Training for Engineers: Adopting Research Ethics Training for Engineering Graduate Students

Sara R. Jordan and Phillip W. Gray

Abstract Two key challenges haunt ethics teaching: relevance and universalism. Demands for relevance, whether relevance is judged based upon association to a local standard (e.g., national professional association code) or to the specific demands of a professional workplace, pull the teaching and study of ethics towards the particular. Calls for a global standard pull ethics teaching and scholarship toward high level principles that can be difficult to justify at a level students and grants funding agencies find applicable. Particularity and specificity complicate demands for global standards or universal norms, while exposition at a universalized level frustrate application in a relevant context. This chapter investigates the structure of research ethics as exemplified in the Singapore Statement, then turns to a case study of RCR students in Hong Kong to provide an example of RCR receptivity among students in engineering and other fields.

Keywords Ethics • Applied ethics • Responsible conduct of research • Engineering ethics • Engineering education

13.1 A Bold Conjecture

A current trend in tertiary ethics training is to narrow the focus of ethics to applied ethics, or ethics for specific disciplines. The popular textbooks for ethics courses make this inclination obvious: there are ethics texts for engineers (Harris et al. 2009), for psychologists (Koocher and Keith-Spiegel 1998), for social scientists (Israel and Hay 2006), for biologists (Laake et al. 2007: 33–82), for clinical science disciplines (Emanuel et al. 2008), and for scientists in general (Committee on

S.R. Jordan (✉)
Virginia Polytechnic Institute and State University, Blacksburg, VA, USA
e-mail: sjordan@vt.edu

P.W. Gray
Texas A&M University at Qatar, Doha, Qatar
e-mail: phillip.gray@qatar.tamu.edu

© Springer International Publishing Switzerland 2015
C. Murphy et al. (eds.), *Engineering Ethics for a Globalized World*, Philosophy of Engineering and Technology 22, DOI 10.1007/978-3-319-18260-5_13

Science, Engineering and Public Policy 1998), among others. Limiting ethical consideration to specific disciplines has great intuitive appeal, insofar as each field has unique conundrums and challenges that may lack direct equivalents in other scholarly or professional areas. The appeal of specification is that it also promotes avoidance of otherwise superfluous topics that might be covered in a general ethics course (i.e., petroleum engineers spending time on ethical quandaries of consent by minors to school-based curriculum research).

Yet, the conventional assumption that "narrow equals better" is contestable—narrowly focused courses may reinforce professional biases, limit the scope of moral imagination, or even perpetuate questionable behaviors that might be openly questioned if presented more generally. In this chapter, we challenge the assumption that domain restricted courses are "better than" broad courses by presenting a case for adaptation of Responsible Conduct of Research teaching for graduate-level engineers needing training in Engineering Ethics.

We propose that Responsible Conduct of Research (RCR) training is an ideal, "local minimum" curriculum that can be expanded to fit a variety of other professional ethics topics. As we see it, RCR principles and practices could be one potential platform upon which to build a global form of engineering ethics training that incorporates universalistic assumptions and moral norms that reach beyond the entrenched codes of practice commonly used as a starting point in a variety of applied ethics courses. Our starting point is that, although various fields of academic and professional study have their own innumerous specific needs, they each work within the overarching paradigm of scientific inquiry for which the basic structure, demands, and practices remain similar. Indeed "research" or, more broadly, "science" serves as the *lingua franca* of exploration, whether of scholarly matters or client-based work or quality improvements and quality assessment even, thus research episteme connects together the varying strains of the disciplines. Given the universal nature of "research" as systematic, generalizable inquiry, RCR training, serves as a basic foundation for engineering ethics courses, whether within or outside of the occidental context.

For example, the basic structure of something like the Singapore Statement provides a useful model of generalizable statements of ethical practice that apply to scientific inquiry. The Singapore Principles of Research Integrity (2010) presents various principles that are applicable to research in all fields, such as honesty, accountability, professional courtesy and fairness, and good stewardship. Honesty is emphasized as an essential intrinsic motivation for practitioners in all aspects of research. Accountability is proposed as a guiding principle for the conduct of research; all researchers should be able to account for each element of their research input and output to an inquiring public of laypersons or fellow professionals. Professional courtesy and fairness advise research professionals to address the needs of personnel management that come with working with others in research endeavors. Finally, good stewardship denotes the importance of attending to research on behalf of others, particularly the lay public for whom research lays the foundations for improvements in infrastructure, health, education, and civil society.

The use of the Singapore Statement on Research Integrity as a backdrop for engineering ethics training is a not a random or arbitrary choice. The development of this statement through extensive "crowd sourcing" with the 300 research professionals from all disciplines and 26 nations around the world attending the Second World Conference on Research Integrity in July 2010 suggests that this statement reflects the desired goals of inclusion of disciplinary relevance and global inclusive purchase. Further, as applies to the specific domain of engineering, the Singapore Statement principles are substantively similar to those pf the American Society of Civil Engineers Code (ASCE 2013) and the IEEE Code of Ethics (2013), among others. More broadly, and potentially of assistance for teaching graduate level engineering courses, for which there is a more strongly interdisciplinary component, the Singapore Statement also overlaps with non-engineering professional codes, including the ICH-GCP Good Clinical Practice standards, the Declaration of Helsinki from the World Medical Association (WMA), and the code for the American Society for Public Administration (APSA), among many others.

In what remains of this chapter, we describe how RCR training was used to teach engineers professional ethics in the context of a multinational university outside of the global west, which served a population of students from developed, developing and emerging nations.

13.2 RCR Training

Responsible Conduct of Research training, which might also be called research ethics or research integrity training, is defined by Steneck as instruction which aims to instill in learners, "The quality of possessing and steadfastly adhering to high moral principles and professional standards, as outlined by professional organizations, research institutions and, when relevant, the government and public" (2006, 56). Others, such as the National Academies and the Institutes of Medicine, have refined Steneck's broad definition to distinguish two types of research integrity—institutional and individual. For example, "for the individual scientist, integrity embodies above all a commitment to intellectual honesty and personal responsibility for one's actions and to a range of practices that characterize responsible research conduct" and at the institutional level, "it [research integrity] is a matter of creating an environment that promotes responsible conduct by embracing standards of excellence, trustworthiness, and lawfulness that inform institutional practices" (2002, 34).

These encompassing definitions of research integrity are often followed up by more or less specific discussion of specific practices that constitute integrity. We attribute this to the relationship between research integrity and professional ethics, within which there is a consistent tendency to enumerate specific practices as definitional for integrity. Within the RCR literature, this is best shown in Macrina's work *Scientific Integrity* (2005, 12), which defines research integrity as one of the four areas of responsible conduct of research practice, which itself is divided into four areas—"matters pertaining to data", "authorship and publication practices", "mentoring", and "collaborative research". Within the book, Macrina deftly elides scientific integrity with research integrity, which he defines further as one of four components of responsible conduct of research. While Macrina's work blends concepts uncomfortably, he offers clear suggestions that scientific practice ought to conform to the norms of "honesty, objectivity, and collegiality" (2005, 15). Adherence to these norms, along with the expected conventions of the scientific method, defines good conduct in scientific practice.

Macrina does not, however, synthesize his suggestions into a single statement as was done in the Singapore Statement. Emphasis on professional practices and standards is a persistent theme in research integrity definitions. For example, the recent Singapore Statement on Research Integrity suggests that the following "principles and professional responsibilities . . . are fundamental to the integrity of research":

> PRINCIPLES: honesty in all aspects of research, accountability in the conduct of research, professional courtesy and fairness in working with others, good stewardship of research on behalf of others; RESPONSIBILITIES: integrity, adherence to regulations, research methods, research records, research findings, authorship, publication acknowledgement, peer review, conflict of interest, public communication, reporting irresponsible research practices, responding to irresponsible research practices, research environments, societal considerations. (2010)

The Singapore Statement synthesized a number of nations' and disciplines' codes of research practice. Importantly, however, the structure of the document points clearly to the historical origin of the various ideas of RCR training previously captured by the US Public Health Service.

Within the 1989 "Requirement for Programs on the Responsible Conduct of Research in National Research Service Award Institutional Training Programs", the NIH established that "producing high quality researchers" requires " . . . attention be directed towards scientific integrity in the conduct of research" (1989, 1). The 1989 document did not outline clearly specific practices. However, within the 1994 "Reminder and Update: Requirement for Instruction in the Responsible Conduct of Research in National Research Service Award Institutional Training Grants", RCR was described as an area of "training" or "instruction" involving "instruction in the following areas: conflict of interest, responsible authorship, policies for handling misconduct, policies regarding the use of human and animal subjects, and data management". The clear connection between concept and practices emerged in the 2009, "Update on the Requirement for Instruction in the Responsible Conduct of Research", wherein "responsible conduct of research is defined as the practice of scientific investigation with integrity. It involves the awareness and application of established professional norms and ethical principles in the performance of all activities related to scientific research" ("Definition"). Despite the vagueness of definitions offered in the earliest years of RCR training (see Steneck and Bugler 2007, 829–831), the enumeration of these and other practices as the definition of RCR is now common to the works of regulators and non-regulators.

As the components of RCR training have evolved to become more expansive, so have the requirements for essential training formats. Under current US regulations, there are various requirements.[1] The format must include face-to-face discussions and faculty participation in both formal (classroom) and informal (graduate supervision) contexts. The documented duration must be at least 8 h of contact time for recipients of relevant grants. In response to some institutions' attempt to streamline or outsource RCR training, the most recent regulations stipulate that on-line training cannot be the sole form of RCR instruction. Instead, in-depth case discussions and experiential learning are recommended. The pressure to standardize RCR training pedagogy arose due, in part, to the inter-institutional, interdisciplinary, and international nature of contemporary research. As institutions holding grants from agencies requiring RCR training sought to ensure compliance with evolving regulations, many of these institutions experimented with various modes of delivery (e.g., online), expanded training to cover individuals not currently under grant obligations, and began to turn the limited scope of RCR training into an institutional imperative.

The creeping evolution of Responsible Conduct of Research (a.k.a., "Ethics Creep" (Haggerty 2004)), has led some observers, such as the present authors, to note systematic deficiencies in RCR training. Some of these deficiencies apply to ethics teaching in general, while others are local to the RCR context. First is the tendency for professional ethics teaching, to include RCR and research ethics teaching, to be wedged into the curriculum. Despite policy exhortations to the contrary, RCR is not commonly or fully integrated throughout a course. Consequently, ethics teaching often becomes an "add-on," thus separating and isolating the ethics training and the topical substance of the course. Second, instead of the formal and informal career span training advocated in US regulations, RCR training becomes a "box to check" before performance reviews. This leads to the troublesome practice of the "half-day workshop" approach to ethics training. Limited in its scope and time, these half-day activities take on an artificial quality, leading not towards ethical learning but instead mere compliance and "checking off" of requirements. As part of a compliance focused type of ethics training, check box RCR is a form of "low road ethics." For critics, RCR training tends towards a "checklist" view of ethics, where "checking the boxes" on what should be done is more important than understanding why this training is important, what it ought to accomplish, and where these principles apply in the daily life of an active researcher. Another criticism that is made consistently is the biomedical focus of RCR training. The emphasis on human participant protections and the extensive use of clinical trials examples, lead those outside of the clinical sciences to question the usefulness of such information in their own disciplines and research. Finally, RCR training too often simply becomes Research Misconduct (RM) training. Too often taught in a "Don'ts, not Dos" approach, RCR training descends into a mere lists of negative instructions. Despite how it is idealized in the regulations, RCR training

---

[1] Although other regulatory regimes may also be relevant, U.S. regulations play an important role in higher education across borders. The policies regarding research integrity found in the US Federal Wide Assurance, for instance, tends towards a globalization of U.S. regulations in the area of research.

becomes "don't plagiarize" and "don't put yourself in a financial conflict of interest situation", instead of encouraging good practices. Finally, the question of audience for ethics training is an important one. If ethics training is to have an effect, it may need to be put in the context of relevant, topical, practice. This means it may be poorly suited to first-year or even early career researchers for whom many of the problems of conflict of interest, for example, may not have arisen yet.

Having noted the general structure, benefits and demerits of RCR training, we must discuss its practice. If RCR training is to serve as a minimal foundation for a global form of engineering ethics training, how does one accentuate the positive aspects of RCR training while avoiding the tendency to fall into its corroded forms? We now turn to a case study of a RCR course. More specifically, this case comes from a tertiary institution in Hong Kong, thus providing a broader, global view to what has thus far been a mostly American-focused discussion.

13.3 The Experience in Hong Kong as Case Study

The Hong Kong Special Administrative Region (HKSAR) presents an interesting example for our purposes. A colony of the United Kingdom until 1997, and now part of People's Republic of China under the "One Country, Two Systems" arrangement, Hong Kong's culture is a mixture of Western and Asian norms.[2] Our case takes place at the University of Hong Kong. In the 2007–2008 year, teaching research ethics was a voluntary matter. But also in this year, a major scandal affected the university: specifically, the Dean of Medicine was indicted for embezzlement and research fraud (he would eventually be jailed for 22 months). Thus, in the 2008–2009 year, teaching in research ethics expanded from a single, elective course in bioethics to a university wide, five sections course on RCR. From the 2009–2010 year and up to the present, there is a compulsory Research Ethics course for all post-graduate students that teaches an RCR curriculum. There is also a half-day Responsible Conduct of Research seminar for faculty, which is elective for all current faculty but recently made obligatory for all new faculty. At present, University of Hong Kong is the only university in the territory (1 of 7) that teaches RCR explicitly and regularly. This university is also only one of two with a regular institutional review board, animal care and use committee, and medical ethics faculty.

13.3.1 RCR Training Course and Engineering Disciplinary Division

As designed by a multidisciplinary group of scholars and led by the primary course instructor, the HKU course lasts for 6 weeks, with 2 hour lectures/case

---

[2]For a fuller discussion, see Jordan with Gray (2013).

analysis (and thus a total of 12 contact hours). This design has four university mandatory components: research misconduct, conflict of interest, human and animal protections, and authorship ethics. In terms of disciplinary divisions there are five: biomedical/clinical; physical/non-clinical sciences; social sciences and education; arts, humanities, and architecture; and engineering. Requirements include a minimum attendance at four of the six lectures, the completion of CITI Program Modules, and an essay examination.

In the engineering disciplinary division, each of the main course components of RCR training is connected to the specific practice of engineering itself, with two additional components as well. For research misconduct, the engineering course focused on falsification and fabrication of data in research and client reports. Authorship ethics concentrated on the issue of group authorship in engineering works. The conflict of interest section focused upon the connections of research and practice, and in particular the issues surrounding academic-industry sponsored partnerships. Closely related to this topic was academic-industry collaboration and the research-practice blend: for the career stage of the class participants, the main theme was working as a post-graduate student on an industry sponsored project and the ethical questions that can arise. Human and animal protections also had a place in the engineering division, noting engineering controls and improving human and animal protections (e.g., redundant protections) are often as much an engineering matter as a matter for disciplinary scientists. Finally, the engineering course noted the issues of data management, particularly concerning large data sets, and stewardship of data, ownership of data, and data sharing in industry-academic collaboratives.

13.3.2 Research into RCR Training: Survey

In the 2009–2010 year, the course instructor was asked to conduct an RCR knowledge and attitudes evaluation among the students in the course. The evaluation was sponsored by the Vice-Chancellor of Research at HKU, an adherent to "evidence based medicine" who had read a text by Atul Gawande (2007), and in that spirit desired to know how to make RCR training better.[3] A pre-post test survey of all students in the course served as a basic attitudinal and knowledge survey. The course instructor used this opportunity to test various hypotheses regarding RCR training, including RCR training in the context of China (Jordan and Gray 2013), as well RCR training and the relationships between post-graduate students and their supervisors (Jordan and Gray 2012; Gray and Jordan 2012). Here, we will consider some hypotheses on the nature of research ethics as it pertains to specific disciplines.

---

[3]This research was funded in part by a grant from the Office of the Vice-Chancellor for Research at the University of Hong Kong.

As noted above, we aver that research serves as common language across disciplinary and professional boundaries, thus leading to commonalities in ethical training. However, research itself is specific to a disciplinary perspective. From this specificity, we derived two testable hypotheses.

1. *Researchers from each discipline will respond to ethics training differently.*
2. *RCR training, being more biomedical specific in its origins, will be more appreciated by students of medicine than by any other discipline.*

These hypotheses examine some of the issues discussed above regarding RCR training. Our first hypothesis addresses the universality (or lack thereof) of RCR training. If research is discipline-specific to the degree that post-graduate students will respond to ethics training differently, then RCR training as a suitable minimum curriculum for a global engineering ethics training (as well as for other disciplines) comes under serious question. The second hypothesis addresses the historical origins of RCR training in biomedicine, noting that such origins would lean medical students to have a greater appreciation of RCR training than any other discipline.

The survey[4] was designed with a fixed-response Likert-scale model, with possible values to select being *strongly agree, agree, disagree,* and *strongly disagree.* "Neutral" was not included as to mitigate some of the likely acquiescent response bias among the students (cf. Smith 2004; Watkins and Cheung 1995). The survey consisted of 30 substantive questions, as well as questions on demographic traits (gender, age, race/ethnicity, MPhil/PhD/other status, and faculty to which a student belonged). In the 2009–2010 course, there was a total population of 640; the number of useable, complete responses was 1,002 (549 on pretest, 453 on posttest) for a total response rate of 78.2 % (pretest crude response 85.7 %; posttest crude response rate 70.7 %).[5]

The survey itself sought student perceptions on numerous areas in research integrity, including the students' (self-perception of) knowledge of research integrity concepts, as well as their beliefs regarding ethical issues (including plagiarism, data integrity, interactions with supervisors, and peer review, among others). To test the hypotheses presented in this chapter, we used two questions from the broader survey. The two questions used were Question 2 (Q2): "I know what research integrity means;" and Question 8 (Q8): "It is important to record accurately all processes and results from research projects." Each question serves a specific purpose. Q2 indicates two things: first, the pretest indicates the level of self-perceived comfort with research integrity as a concept among post-graduate students in different disciplines; second, the posttest gives indications about what effects, if any, RCR training had on these self-perceptions. Q8 was chosen because of its

---

[4]The Human Subjects Ethics Committee for Non-Clinical Faculties of the University of Hong Kong approved this project under an expedited review with a waiver of documentation of written informed consent. The full survey instrument is reprinted in Gray and Jordan 2012: 309–310.

[5]For an extended discussion of the methodology, cf. Jordan and Gray 2012: 301–303; Gray and Jordan 2012: 301–302.

Table 13.1 Q2: I know what research integrity means

	Strongly agree (%)	Agree (%)	Disagree (%)	Strongly disagree (%)
Pretest				
Engineering	6	54	32	1
Medicine	12	61	25	1
Social science	15	64	17	2
Posttest				
Engineering	27	69	0.9	1
Medicine	22	75	1	0
Social science	25	74	0	0

Table 13.2 Q8: It is important to record accurately all processes and results from research projects

	Strongly agree (%)	Agree (%)	Disagree (%)	Strongly disagree (%)
Pretest				
Engineering	53	39	4	0.9
Medicine	62	34	1	0.7
Social science	51	48	0	0
Posttest				
Engineering	43	54	0.9	0.9
Medicine	62	36	0	0.8
Social science	40	59	0	0

particular relevance to engineering post-graduate students: of all aspects of research ethics, presumably the requirement to record accurately the processes and results of research would be highly salient for them. As such, engineering responses in this area are of particular interest in relation to RCR training to disciplines outside of the clinical sciences. For comparative purposes, we looked at the results from students classified as in engineering, in medicine, and in the social sciences. We will consider each result in turn (Table 13.1).

On Hypothesis 1, we have a null finding. On the pretest, engineering students did differ from both medical and social science students, reporting somewhat lower responses to "strongly agree" and "agree", and somewhat higher in "disagree", on self-perceived knowledge of research integrity. In the posttest run, however, these differences effectively vanished. Engineering students who indicated strong agreement (27 %) varied from medicine students (22 %) and social science students (25 %) to a minor degree. This minor difference also held in the other options as well (Table 13.2).

The findings for Hypothesis 2 are more complex. On accurately recording processes and results, students in medicine effectively did not change between the pretest and the posttest. However, we did see an expected change in both engineering students and in social science students: specifically, we see a shift of students moving from "strongly agree" to "agree." In the case of engineering students in the pretest, 53 % strongly agreed and 39 % agreed. But in the posttest, only 43 %

strongly agreed and 54 % agreed. One partial explanation for the shift could be the movement of the 4 % of pretest students who selected "disagree" possibly shifting to one of the "agree" fields. The difference, however, is notable. A similar change occurs with social science students, but without the same partial explanation (in both pretest and posttest, none of the students selected the disagree options). This finding, therefore, is curious. While it does show perhaps an increased skepticism on the part of students after completing the RCR training, it does not show a strong shift: in other words, students are not disagreeing that accurate record keeping is important, but merely doing so with less strength. We have observed similar post-course trends in responses to supervisor ethicality in other studies (Jordan and Gray 2012: 309–311; Gray and Jordan 2012: 306–308). As we noted there,

> As RCR training informs and sensitizes students to the requirements of good research conduct, they may be less willing to assume quiescently that their supervisors act in ethical ways. Instead, the information acquired from RCR training may create a healthy level of skepticism and critical awareness among students that is reflected as this cooling of trust. (Jordan and Gray 2012: 310)

This increased critical awareness from RCR training may have a similar effect more broadly. Indeed, it appears to have the counterintuitive result that medical students – those we would assume are the most inclined towards RCR training because of its background in the clinical sciences – are unaffected by increases in critical awareness as compared to their compatriots in engineering and social science. One must not lose sight of the major commonality between all disciplines examined, however. For all three, they ranked similarly in terms of agreement against disagreement on the pretest (engineering: 92 %; medicine 96 %; social science 99 %) as well as the posttest (engineering 97 %; medicine 98 %; social science 99 %).

Coming back to the hypotheses, our results testify to null findings for both. It appears that RCR training therefore is working, and is not limited in its usefulness solely to one discipline or specific form of research. To better assess the training's value, we now turn to student evaluations of the course itself.

13.3.3 Student Comments on RCR Training Course

Beyond the survey seeking knowledge and attitudinal data on the students, we also can offer information presented via student feedback on the course itself. This feedback was gathered through regular student evaluations of teaching conducted at the end of the semester, as well as through open comments. Students considered the best thing about the class to be the teacher herself, specifically her enthusiasm and examples. For example, in response to the evaluation question, "What was the best part of this course?", one student responded, "The Teacher! I learned to be a more funny teacher from Dr. Sara [sic]". This indicates that instructor engagement and energy play a significant role in engaging students who, all things being equal, are not inclined to take an ethics course without inducement.

The area students wished to see improved pointed to making the course more interactive. Additionally, students wanted to spend less time on human subjects/animal subjects, focus more on relevant cases, and spend more time reviewing cases in class. Students also complained about the length of cases to be read. One student quipped, "I have no time for this reading! Four pages for each case is too much". This emphasis on cases and case studies repeatedly arises among engineering students, and is reflective of the constant use of case studies as a pedagogical tool in engineering courses of various types.

A final point was the need for a textbook, and more specifically, a textbook that would also be English as a Second Language (ESL) friendly. The textbook itself would include cases, case questions, examples, and resources for further learning, but would also avoid being overly US- or Western-centric, both in its examples as well as in its language. The student's comments as well as those from other faculty were incorporated into the development of a subsequent university textbook, which is available on-line here: http://www.med.hku.hk/v1/research/research-ethics/other-informationreferences/

These comments illustrate the delicate balance an instructor must maintain in providing ethics training to engineering students, where instructor enthusiasm, interaction, and use of case studies are key elements in enhancing student attention (and thus, hopefully learning). As a matter of pedagogical methodology, the engineering ethics instructor is constrained to rely upon case studies more than may be preferred, particularly for instructors who do not come from an engineering background. The instructor must therefore find the balance between abandoning case studies – with students thus required to learn significantly new materials via an unfamiliar teaching method – while also not falling into the trap of "teaching to the case." This challenge is exemplified in enhancing interaction within the class. On the one hand, engineering students are accustomed to hands-on styles of teaching, and thus integrating more student interaction via case study analysis or even role-playing a case would be highly beneficial. But on the other hand, the forms of conceptualization and reasoning in ethics (research, engineering, or otherwise) are significantly different from the versions found within engineering: thus, too much reliance on interaction via case studies could lead students to address the issues presented as design problems instead of ethical problems. For a well-rounded course, the mixture of interaction and instruction is necessary, but can also be taxing.

The final part of this balance, and the one least amendable to curricular rigor, is the role of the instructor him/herself. For engineering students – as well as students in numerous other empirical disciplines – such an ethics course is likely the first or at best second course they will encounter where ethical concepts and reasoning are the primary and substantive focus. Anecdotally, it is recognized that student receptivity to a new topic often depends significantly on the teaching style, competence, and enthusiasm of the instructor. An active and engaged lecturer who lacks understanding about basic engineering concepts will be problematic, as will a professional engineer who merely reads from the textbook. Below, we provide some suggestions to increase the likelihood of choosing engineering ethics instructors who can attain the necessary balance for student needs.

13.4   Recommendations for RCR and Engineering Ethics

Proceeding from the discussion above, we can now make some recommendations for RCR training as a basis for engineering ethics curriculum. We present three main recommendations. RCR and engineering ethics courses could mutually benefit from a repository of relevant cases of research misconduct and research integrity in engineering disciplines. Second, acknowledgement that the scientific paradigm is the background of engineering and other research disciplines would help to situate the need for a shared ethic. Finally, both RCR and engineering ethics training could benefit from acknowledgement of the cross cultural complexities of the modern practice of each. We will discuss each of these recommendations in turn.

As the dominant pedagogical style in engineering, an engineering ethics course based in RCR-training should base itself on case based teaching. These cases do not have to be specifically about engineering, but they should be relevant. The importance of examples should not be undervalued, in particular those examples that are spontaneously delivered. The danger in case-based learning is "learning to the example" – instead of ascertaining the more fundamental ethical issues in case, students may instead simply regurgitate the mere "lessons" of a particular incident. Examples delivered in an extemporaneous manner, therefore can counteract this inclination towards rote repetition by broadening the base of examples students can consider. Although the instructor in such a course need not be an engineering specialist, such a background would be conducive in providing such "off-the-cuff" examples.

The focus upon RCR training as a minimum curriculum for engineering ethics courses should not lead us to fall into the narrow notion of training discussed at the beginning of this chapter. It is not a necessity for such RCR training to be disciplinary specific. It could be paradigmatic instead. An emphasis on the ethical commitments of an engineering philosophy of science or the philosophy of pragmatism broadly may provide some fertile ground for bringing a paradigmatic focus for RCR and engineering ethics.

Finally, a cross-cultural focus would be beneficial for both areas. However, cross cultural could mean many things in the area of engineering ethics: it could be cross national cultures or cross disciplinary cultures. As there remains a spectre of ethnic essentialism in cross cultural training, it may be most beneficial and least offensive to focus on cross-disciplinary blending in courses. For instance, mixed classes may help to improve the "uptake" by students of critical ethical awareness and learning via examples outside of their fields. Students within a discipline will usually know, at least at a vague manner, about specific misconduct events concerning their field and the conventional responses to them: examples include the Milgram experiment in psychology; the Tuskegee syphilis experiments in medicine; or the Kansas City Hyatt Regency walkway collapse for engineering. Before reaching an ethics class, they may "know" the "right" answer through references in their previous classes, despite not knowing *why*. Being confronted with examples outside of their discipline will require students to use their critical reasoning skills and ethical training rather than only recite the usual answers from within their own field.

Putting these recommendations into practice will both enhance engineering ethics education as well as place it within the larger interdisciplinary discourse of research integrity. But these recommendations are only a starting point for advancing the combination of responsible conduct of research training and engineering ethics education. There are additional issues and questions to face in the realms of research and teaching in ethics education, some of which we turn to now.

13.5 Future Directions in Research/Teaching

Designing disciplinary specific courses may help to improve student evaluations of RCR teaching and may improve uptake of the basic principles taught. However, a central question for RCR research is: does RCR education "work" long term? Does RCR education actually advance research integrity? And, how would we measure integrity in research situations? At an individual or institutional level? In asking if RCR training works in the long term, future research should also look to the specific parts of RCR training. For instance, do students who have been trained in RCR engage in misconduct less than those who have not received RCR training? Is there a greater level of voluntary participation in research compliance (e.g., IRB applications) among those trained? On an academic side, should RCR training be "more ethical" instead of compliance-oriented? All of these questions speak to the relevance of RCR training in general. However, we stand by our original "bold conjecture" that RCR training is the most suitable basis of graduate engineering ethics training.

On a practical level, such research would involve re-running the RCR-engineering class survey. While the HKU survey provides useful information, running the survey in multiple locations would help us distinguish similarities and differences not only across disciplines, but also across cultural borders.[6] The additional data would also aid in creating a dataset large enough to run robust statistical analyses, and thus better ascertain correlated interactions between various themes. A similar survey analysis, separate from the previous survey, would ask students about their beliefs regarding ethics in their professional or research practice. The first survey aimed at knowledge and attitudinal data from students on research *integrity*; this survey would instead attempt to ascertain student attitudes and beliefs on research *practice*. Another area of research would be to track rates of falsification, fabrication, and plagiarism (FFP) in dissertations and theses among post-graduate students in the classes taught.

In the area of teaching and pedagogy, there is a significant need for an RCR casebook for engineers. Examples of RCR often come from the fields of clinical sciences and those dealing with human and animal subjects. A casebook of examples

---

[6]For an examination of different cultural norms as they influence professional ethics – in this case, the civil service – cf. Jordan and Gray 2011.

specifically focused upon RCR issues of engineers would advance such training significantly. Additionally, such a text could serve as the basis for a "teaching case bank" of engineering-specific RCR examples, accessible online for instructors when creating and teaching such courses. These teaching aids are useful for creating a well-structured and substantive RCR-engineering course. But what is needed most of all is engineers with RCR interest. Instructors of RCR training often hail from the clinical sciences or the social sciences, usually because it is these scholars who take a strong interest in RCR. This does not imply that engineers do not care about research integrity: rather, it indicates that engineers instead often come at such instruction with an interest in more generalized forms of ethical examination rather than a specifically RCR focus. In creating an academic and professional culture that centers on responsible conduct of research, it is engaged interest of engineers in the academy and in the field that will make the difference.

13.6 Back to the Bold Conjecture

In conclusion, let us return to the beginning and our conjecture. Can RCR training be used for a global engineering code and training plan? Indeed it can, but with reasonable reservations about the insufficient evidence for RCR effectiveness already accounted for.

The ethical principles involved in RCR are mid-level, and thus useful for a practical discipline like engineering. Its principles are not so abstract and "high" – formulated in a manner like Kant's "categorical imperative" – that they present little applied guidance for engineers operating within a research group at a university or corporation. Similarly, it is not so detailed and "low" that it prevents generalizability: in other words, it is not so detailed that its ethical guidance can be summed up as "how to prevent a walkway at the Hyatt Regency Hotel in Kansas City from collapsing in 1981." RCR training, then, presents a happy balance between theory and practice that is particularly conducive to the engineering discipline.

Beginning from a foundation in the practice of research – applicable to numerous disciplines and professions – RCR training provides a backdrop for engineering ethics training that permits communication across both disciplinary and cultural boundaries. Within any given profession, there are exemplar cases that garner particular attention, such as the Milgram experiments in psychology or the Hyatt Regency Hotel case in engineering: by starting from a broad groundwork of research ethics, rather than from any particular disciplinary area, each of these exemplar cases can provide insights and assistance to students across professions. As we have discussed the benefits in cross-discipline training and interaction, we now briefly turn to its importance cross-culturally.

One of the great challenges in professional ethics training, and particularly for engineering ethics, is the issue of relevance to the particular context and environment in which professionals will operate. Typical cases in engineering ethics training often encounter this predicament: although useful as an example of whistleblowing in the face of catastrophe, Roger Boisjoly in the case of the 1986 *Challenger* explosion has less salience for engineers not operating in a national context of space exploration. More importantly, the significance of the *Challenger* case is usually based upon social saliency, insofar as the audience is American and thus has a cultural familiarity (or may have even watched the explosion on live television) with the incident. For those outside of the American context, the Challenger case still proves interesting, but lacks the type of visceral force that can be anticipated in an American audience.

The need, then, is for cases from a multitude of national and cultural settings. By basing engineering ethics training upon a foundation of RCR and research ethics, the case "dataset" increases significantly. Cases from students' own disciplinary backgrounds (engineering, medicine, anthropology, and others) can be examined, illustrating to students both the expanse of ethical issues that professionals face, while also demonstrating the significant overlap in ethical quandaries (and their solutions) that can be discovered across disciplinary boundaries. But another significant benefit of this RCR extension is the ability to address different regional levels of research and engineering development. The problems of the most wealthy and research-intensive nations are not the same as those in emerging economies and developing engineering markets. Looking to the example of Qatar, the difficulties presented by *wasta* for discussions of conflict-of-interest have few useful analogies in cases from the United States. However, the custom of *wasta* share marked similarities to the practice of *guanxi* within parts of China, leading to similar problems with conflict-of-interest. Starting ethics training from an RCR perspective can illuminate these types of similarities across cultural traditions, and allow for additional focus on ethical issues that present greater challenges in developing economies (that also have different or less rigorous regulatory environments) than those usually examined within professional ethics courses. Starting from a non-national and non-disciplinary foundation in RCR, engineering ethics training can better integrate examinations of culturally similar practices from significantly dissimilar societies.

The engineering profession has long been global in its connection with international commerce. Increasingly, however, engineering is also globalized in its nature, with innovations, research, and best practices coming from across the world. For ethical practice, we need a common "language" that permits a Qatari engineer to communicate clearly with engineering colleagues from France, Japan, and Brazil, and that also assists ethical discussions between engineers, biologists, clinical scientists, physicists, organizational psychologists, and other practitioners in expanding human knowledge. The first step to replacing the veritable cultural/disciplinary Tower of Babel with unified "language," supplanting cacophony with coherence, is via a common basis in ethical understanding. For engineering ethics, and professional ethics in general, Responsible Conduct of Research presents one "local minimum" that could be a global way forward.

References

American Society of Civil Engineers (2013). Code of ethics. Available at: http://www.asce.org/Ethics/Code-of-Ethics/. Accessed 3 July 2013.

Committee on Science, Engineering and Public Policy. (1998). *On being a scientist: Responsible conduct of research* (2nd ed.). Washington, DC: National Academic Press.

Emanuel, E. J., Grady, C., Crouch, R. A., Lie, R. K., Miller, F. G., & Wendler, D. (Eds.). (2008). *The Oxford textbook of clinical research ethics*. Oxford: Oxford University Press.

Gawande, A. (2007). *Better: A surgeon's notes on performance*. New York: Picador.

Gray, P. W., & Jordan, S. R. (2012). Supervisors and academic integrity: Supervisors as exemplars and mentors. *Journal of Academic Ethics, 10*(4), 299–311.

Haggerty, K. D. (2004). Ethics creep: Governing social science research in the name of ethics. *Qualitative Sociology, 27*(4), 391–414.

Harris, C. E., Pritchard, M. S., & Rabins, M. J. (2009). *Engineering ethics: Concepts and cases* (4th ed.). Belmont: Wadsworth.

IEEE. (2013). IEEE code of ethics. http://www.ieee.org/about/corporate/governance/p7-8.html. Accessed 3 July 2013.

Israel, M., & Hay, I. (2006). *Research ethics for social scientists*. London: Sage.

Jordan, S. R., & Gray, P. W. (2011). *The ethics of public administration: The challenges of global governance*. Waco: Baylor University Press.

Jordan, S. R., & Gray, P. W. (2012). Responsible conduct of research training and trust between research post-graduate students and supervisors. *Ethics & Behavior, 22*(4), 297–314.

Jordan, S. R., & Gray, P. W. (2013). Research integrity in Greater China: Surveying regulations, perceptions and knowledge of research integrity from a Hong Kong perspective. *Developing World Bioethics, 13*(3), 125–137.

Koocher, G. P., & Keith-Spiegel, P. (1998). *Ethics in psychology: Professional standards and cases* (2nd ed.). Oxford: Oxford University Press.

Laake, P., Benestad, H. B., & Olsen, B. R. (Eds.). (2007). *Research methodology in the medical and biological sciences*. London: Academic.

Smith, P. B. (2004). Acquiescent response bias as an aspect of cultural communication style. *Journal of Cross-Cultural Psychology, 35*(1), 50–61.

Steneck, N. H., & Bugler, R. E. (2007). The history, purpose, and future of instruction in the responsible conduct of research. *Academic Medicine, 82*(9), 829–834.

Watkins, D., & Cheung, S. (1995). Culture, gender, and response bias: An analysis of response to the self-description questionnaire. *Journal of Cross-Cultural Psychology, 26*(5), 490–504.

Chapter 14
Training Engineers in Moral Imagination for Global Contexts

William J. Frey

Abstract What challenges do students face in preparing to practice responsible engineering in a global world? How can students be brought to recognize and avoid pitfalls like paternalism? One answer to this question comes from an unlikely source. While satirizing telescopic philanthropy, Charles Dickens lays down the conditions that a curriculum in global engineering ethics must address. One must "adapt [one's] mind to those very differently situated," address other cultures "from suitable points of view," cultivate a "delicate knowledge of the heart," and realize that "good intentions alone" are not enough.

This essay situates teaching engineering ethics in a global context by outlining the pitfalls students must learn to avoid, putting forward moral imagination as a means to recognize and work around these pitfalls, and showing how moral imagination informs the different modes of engineering moral expertise. This paper will conclude with three learning modules that address some of the pedagogical challenges posed by moral imagination.

Keywords Moral imagination • Engineering ethics • Global ethics • Teaching ethics

14.1 Introduction

Moral imagination is not new to the university curriculum. In *Ethics Teaching in Higher Education*, Daniel Callahan sets forth what have come to be known as the Hastings Center objectives for teaching ethics (Callahan 1980, pp. 61–80). "Stimulating the moral imagination" is the first of five goals. Roughly a decade later, Mark Johnson defined moral imagination in terms of a series of advances

W.J. Frey (✉)
College of Business Administration, Call Box 9000 University of Puerto Rico at Mayagüez, Mayagüez, Puerto Rico, 00681-9000
e-mail: williamjoseph.frey@upr.edu

© Springer International Publishing Switzerland 2015
C. Murphy et al. (eds.), *Engineering Ethics for a Globalized World*, Philosophy of Engineering and Technology 22, DOI 10.1007/978-3-319-18260-5_14

in cognitive science: the "theory of prototypes," "frame semantics," "metaphorical understanding," "basic level experience," and "narrative" (Johnson 1994, p. 11). Patricia Werhane characterized moral imagination as

> the ability in particular circumstances to discover and evaluate possibilities not merely determined by that circumstance or limited by its operative mental models, or merely framed by a set of rules or rule-governed concerns. In managerial decision-making, moral imagination entails perceiving norms, social roles, and relationships entwined in any situation. Developing moral imagination involves heightened awareness of contextual moral dilemmas and their mental models, the ability to envision and evaluate new mental models that create new possibilities, and the capability to reframe the dilemma and create new solutions in ways that are novel, economically viable and morally justifiable (Werhane 1999, p. 93).

Both Johnson and Werhane tie their treatments of moral imagination closely to Kant, especially his division of imagination into the reproductive, productive, and free reflection functions (Werhane 1999, pp. 96–107 and Johnson 1986, pp. 139–172). But they also incorporate recent findings from moral psychology, cognitive science, and systems theory.

Moral imagination can be taught, and its teaching should come to be an essential part of a globally-oriented engineering curriculum. For Johnson, moral imagination, like artistic sensitivity, requires a skills-based pedagogy grounded in practice and training.

> As in art, so also in morality, there is a dimension of skill. There are aspects of morality where practice and training are appropriate. Such practice is seldom a case of following verbal rules, though there is a place for this kind of teaching, especially for young children. The skillful coping that is required is the kind of understanding that allows some people to work their way creatively and constructively through situations that are developing as they confront them. This kind of skill, then, is never merely a fixed procedure that one applies mechanically to situations; instead, it is an elusive kind of knowledge of how to go on, in the midst of contingencies and unforeseen circumstances, to realize well-being. It is this form of creative making that I have in mind when I speak of 'orchestrating' relationships or 'composing' situations. (Johnson 1994, p. 214)

Of course, moral imagination is not the only skill required by moral expertise. The Hastings Center goes on to outline four further and quite distinct objectives in ethics in higher education. But there are good reasons to include moral imagination in a curriculum devoted to global engineering. First, engineering practiced in the global arena stands subject to certain pitfalls that Charles Dickens satirizes as "telescopic philanthropy." Moral imagination provides the proper antidote to these as will be seen below. Second, moral imagination is effectively approached through the teaching of three skill sets: role-taking, multiple framing, and dramatic rehearsals.[1] These skills can be practiced and refined through the redeployment

---

[1] Werhane highlights the importance of *role-taking* (and empathy) through her interpretation of Adam Smith's impartial spectator as a construct wrought by moral imagination. See Sherman (1998), pp. 85–96 for another discussion of empathy in Adam Smith. Werhane also works closely with *multiple framing* in her recent book *Alleviating Poverty through Profitable Partnerships* by emphasizing how breaking out of certain habitual modes of framing (mental models or mind sets)

of methods already familiar in many ethics courses. This paper will outline these skills and then show how they are deployed in different forms of engineering moral agency. Third, the last part of this paper profiles three learning activities that effectively address the pedagogical challenges of moral imagination. These activities have been developed through the UPRM's (University of Puerto Rico at Mayagücz) initiative, GREAT IDEA. Such module profiles hopefully will provide the reader with concrete examples of a "practice and training" pedagogy in moral imagination that can form a part of a globally-oriented engineering curriculum.

14.2 Four Cautions to Would-Be Telescopic Philanthropists

In *Bleak House*, Charles Dickens distinguishes "telescopic philanthropy" from a "circle of duty" approach. Esther Summerson (from the circle of duty side) attends to those around her with empathy and compassion. She looks to immediate surroundings, gives importance to concrete matters, addresses needs, and solves problems. Her skill lies in uncovering moral salience in her circumstances. This sensitivity makes it possible for her to help her guardian, John Jarndyce. He wants to help others but lacks her sensitivity to surroundings; acting on his own, he is continually "floored" when his well-intentioned efforts go awry. Esther has opened herself to what is around her and works "to let that circle of duty gradually and naturally expand itself" (Dickens 1852/1985, p. 98).

Dickens satirizes another character, Mrs. Jellyby, as a "telescopic philanthropist." Like a telescope, Jellyby's attention passes over the immediate and concentrates on the distant; Jellyby obsesses on the plight of the natives of the remote African nation, Borrioboola-Gha, to the neglect of her immediate surroundings.

When Esther is pressed by Mrs. Pardiggle (yet another telescopic philanthropist) to join a philanthropical crusade, she declines for the following reasons:

> That I was inexperienced in the art of adapting my mind to minds very differently situated, and addressing them from suitable points of view. That I had not that delicate knowledge of the heart which must be essential to such a work. That I had much to learn myself, before I could teach others, and that I could not confide in my good intentions alone. For these reasons, I thought it best to be as useful as I could, and to render what kind of services I could to those immediately about me; and to try to let that circle of duty gradually and naturally expand itself (Dickens 1852/1985, pp. 97–98).

Esther's reasons pose challenges that can be addressed by moral imagination. Thus, engineers who would address their efforts toward helping those in developing nations should be mindful of the following:

Of adapting their minds to minds very differently situated. It would help globally-focused engineers to learn to see beyond restrictive mindsets like paternalism

---

is necessary to unlock creative possibilities in developing partnerships to attack global poverty (Werhane et al., pp. 46–58). Johnson, influenced by John Dewey, places considerable importance on *dramatic rehearsals* in carrying out moral deliberation (Johnson 1994, p. 149).

(Werhane et al. 2010, pp. 46–58). Instead of viewing individuals in developing communities as needy and deficient, they must recognize their skills, knowledge, and resources. The NGO, Aprovecho, has formulated a useful way of approaching cultures with differently situated minds. Instead of focusing on imparting technical know-how to the "needy" or "underprivileged," they first work to learn what these communities themselves have to offer. They call this way of adapting to the minds of others an *"inverse peace corps"*:

> We wanted to work as an inverse Peace Corps," Ianto Evans, one of the founding members, told me....We would bring in villagers from Kenya or Lesotho, have them stay with us, and teach us what they knew—everything from cooking to growing things to assessing how much is too much. (Bilger 2009, p. 88).

Aprovecho and the UPRM have had several encounters beginning with an "Alternative Job-Fair" held September 27, 2012. In a very frank assessment of their successes and failures, Aprovecho representatives have discussed their struggle to combine good engineering with responsiveness to the cooking practices and traditions of those who would use their stoves. The inverse peace corps presents an important clue to their success; Aprovecho has approached people and groups in the developing world with open minds and with the recognition that learning goes both ways. Engineers who would adapt their minds to those "differently situated" must first recognize that other communities are repositories of valuable traditions, skills, and knowledge.

Of learning to address other cultures from suitable points of view. Telescopic philanthropy passes over the particulars situated on both sides of the telescope. To avoid this, the engineer should cultivate what Harris calls "techno-socio sensitivity," the *"critical awareness of the way technology affects society and the way social forces in turn affect the evolution of technology"* (Harris 2008, p. 16 and Huff 2001). Socio-technical system (STS) description underpins this sensitivity and the success of many engineering projects depends on the ability of planners to factor in the socio-technical system into which the project will be instantiated. In commenting on difficulties surrounding the One Laptop Per Child project, Kraemer et al. argue that "innovators must consider the need for expertise in sociology, anthropology, public policy, and economics as well as for engineers" (Kraemer et al., p. 72). A curriculum exposing engineering students to socio-technical system study and analysis would accustom them to taking this interdisciplinary approach to the communities they would work in. Constricting the curriculum to a narrowly technical approach condemns engineers to miss those aspects of the STSs that led OLPC engineers astray (Kullman and Lee 2012, pp. 45–47).

Of cultivating that delicate knowledge of the heart. Cultivated emotions tune moral agents into their surroundings by focusing attention, signaling value, revealing moral salience, and motivating responsive action (Sherman 1997, pp. 40–50). One cultivates *empathy* by building onto in-born, mimetic capacities (Sherman 1998, pp. 85–96). *Compassion*, opens agents to the suffering of others (Nussbaum 2001, pp. 304–327); *hope* motivates restorative action (Urban-Walker 2006, pp. 44–49). Such emotions (and their attuned sensitivities) cannot be created out of nothing, and no engineering curriculum that tries to do this would succeed. But an impressive list of thinkers, Mark Johnson and Martha Nussbaum among them, argue that these emotions can be attuned and refined in the classroom through the exercise of moral imagination. In this passage, Johnson makes use of John Gardner's concept of "moral fiction:"

> John Gardner has argued that fiction is a laboratory in which we can explore in imagination the probable implications of people's character and choices. He describes what he calls "moral fiction" as a "philosophical method" in which art "controls the argument and gives it its rigor, forces the writer to intense yet dispassionate and unprejudiced watchfulness, drives him—in ways abstract logic cannot match—to unexpected discoveries and, frequently, a change of mind." (Johnson, p. 197; Gardner 1978, p. 108).

Literature invites readers to enter into a narrative context, take a participatory perspective, and experience imaginatively what the different characters experience all the way down to their emotional responses. Engineers need not read Faulkner to develop moral imagination or to practice engineering effectively in globally remote contexts. Often the same kind of exploration, through narrative, character, emotion, and situation can take place by discussing and acting out scenarios and cases that depict global engineering situations. This creates the context for a "dramatic rehearsal" (more on this below) that allows for the exploration and refinement of emotions as many instructors of engineering ethics have already discovered.[2]

Of not confiding in good intentions alone. So many times, technical artifacts are brought to developing countries with the good intention of alleviating the difficulties of day-to-day living. Playpumps were distributed throughout Africa to relieve women from having to walk long distances for water (Smith 2005, pp. 49–52). Installed in school yards, they were designed to provide children with a playground toy that also helped pump water out of the ground and into a storage tank. "The design has an innovative way of converting the circular motion of the toy into up-and-down motion for drawing water using only two moving parts" (Smith 2005, pp. 49–50). Nevertheless, a project set forth with the best of intentions and based on solid engineering went awry. Many Playpumps now stand unused because no one knows how to repair them, and because many villagers prefer to use simpler, hand-operated pumps (Costello 2010). Developers consulted with local communities and had their buy-in before the installation of the pumps. But good intentioned projects need to be supplemented with careful, follow-up study; a technical artifact's operational history often reveals problems that are not manifest in planning, implementation, or early use.

---

[2] Baillie et al. (2010) set up an interesting role play where students identify with participants in the Waste for Life project under development in Buenos Aires, Argentina (Baillie et al. 2010, p. 102). Bilbao et al. (2006), authors of a Spanish textbook on Engineering Ethics, provide a series of cases that require students to take on a participatory perspective in an imagined scenario. In one, students are asked to role play with an engineering student who is being interviewed by an aerospace company that is probing his views on weaponizing space (Bilbao et al. 2009, p. 258).

14.3   Moral Imagination and Its Skill-Sets

Esther's criteria highlight the problem for which moral imagination provides the solution. Moral imagination will help engineers adapt to complex global circumstances. Moreover, it brings into relief the pitfalls of a well-intentioned telescopic philanthropy. To teach it, one must target three skill sets:

"Role-taking" or empathic projection. Moral imagination requires skill in role-taking, i.e., projecting into the perspective of another and viewing a situation from this standpoint. Twelve-year-old, Paul (interviewed by Pritchard), puts it nicely: "Well, it's like your brain has to leave your head and go into the other guy's head and then come back into your head; but you still see it like it was in the other guy's head and then you decide that way" (Pritchard 1996, p. 157). Role-taking, thus, involves empathic projection. But this projection is subject to pitfalls outlined by Nancy Sherman (reacting to Adrian Piper):

> There are the dangers of egocentric projection, but also . . . the converse danger of excessive appropriation of an object such that one erodes a critical stance toward it and the capacity to act from a separate and sympathetic agency. [T]he scylla and charibdis of empathic imagination are self-absorption (i.e., projection) but, equally, vicarious possession (i.e., emotional identification or enmeshment) (Sherman 1998, p. 110).

Modifying Paul's observation, one's brain goes into another's body but then returns with the gathered experience of the other. This, then, is placed on a broader canvas consisting of one's own experiences as well as those of others involved in the situation. For example, UPRM students have successfully projected into the standpoint of Amish people in a module entitled, "Responsible Choice for Appropriate Technology."[3] Such exercises build on the skills students already have in role-taking. The unquestioned assumption these skills must be built out of nothing and that such a task is impossible in the university classroom is a major block to integrating moral imagination into the curriculum. This all or nothing approach misses the fact that teaching strategies such as case discussion and role playing build on existing skills and allow students to practice role-taking, receive timely feedback, and then refine this feedback into further practice sessions (Huff et al. 2008).

Viewing a situation under multiple frames. Framing allows individuals to make sense of their surroundings by giving them structure. Humans only attend to a limited number of things at a time; framing helps select which elements to bring into the foreground and which to push into the background (James 1890/1950, pp. 284–290). Most of the time, this filtering is useful.

---

[3]They read Jamison Wetmore's excellent paper, "Amish Technology: Reinforcing Values and Building Community" and then prepared a poster presentation on one of the instances of technology choice his article outlines. Wetmore's concrete description of the Amish STS helps students go beyond an external and judgmental description to understand how Amish technology choice makes use of shared values, community identity, and common concerns to preserve religious beliefs.

However, framing activities can harden into rigid habits (Werhane et al. 2010, p. 46). The proper antidote is "transperspectivity" (Winter 1990).[4] First, one *"unravel[s] or trace[s] back the strands by which our constructions weave our world together."* This recovers previous framings and brings them to self-awareness so that agents can see how their current views arise from habitual framings. Next, one imagines *"how the world might be constructed differently"* (Johnson 1994, p. 241.) By playing over different possible framings of a situation, agents unlock new possibilities. Patricia Werhane provides a useful list of mental models or frames that block the formation of effective partnerships for alleviating poverty in the developing world (Werhane et al. 2010, pp. 46–58).[5] Moral imagination works against these obstructive mental models; it helps agents to uncover restrictive frames, to bring them out into the open for questioning, and to experiment with different and new modes of framing.[6]

Dramatic Rehearsals. Moral imagination can also create a space in deliberation between an impulse or action plan and its execution in the real world. Dewey shows how this imaginative space facilitates "dramatic rehearsals" that test alternatives before committing them to action in the real world. Here is how he describes dramatic rehearsal in *Theory of the Moral Life*:

> Deliberation is actually an imaginative rehearsal of various courses of conduct. We give way, *in our mind,* to some impulse; we try, *in our mind,* some plan. Following its career through various steps, we find ourselves in imagination in the presence of the consequences that would follow: and as we then like and approve, or dislike and disapprove, these consequences, we find the original impulse or plan good or bad. Deliberation [becomes] dramatic and active.... (Dewey 1960, p. 135)

There are four components to dramatic rehearsals. First, they are *experimental.* Recalling past experience and using this to structure the imagined context of action, dramatic rehearsals allow one to test action alternatives as one would test an experimental hypothesis. (Think again on Gardner's laboratory where moral fiction empowers the imaginative exploration of character and choice.) Second, dramatic rehearsals take on *narrative form.*[7] Playing out an action alternative through a dramatic rehearsal helps one envision how the action might unfold were it enacted

---

[4] Johnson treats Winter's notion of transperspectivity as an essential component of moral imagination (Johnson 1994, p. 241).

[5] Some examples that are useful for engineering students to study: paternalism, the biases of common sense and conceptualism, the research paradox, and "one size fits all." (Werhane et al. 2010, pp. 50–58).

[6] Another exercise in multiple framing takes place according to what Werhane terms the Rashomon effect" (Werhane 1999, pp. 68–88). Rios et al. advocate deploying Rashomon cases where a series of narratives are put forth each from the limited and biased perspective of a participant in the case. Because an overriding, privileged narrative is withheld, students have an opportunity to practice working through different and conflicting ways of framing a series of events.

[7] Narratives provide a particularly effective way to explore emotion. As Urban-Walker puts it, emotions, themselves, are "narratively structured episodes of thought, feelings bodily changes, and expressive activity" (Urban-Walker 2006, p. 49).

in the real world. Third, dramatic rehearsals are built out of and stem from *value sensitivities*. One discovers the values embodied in an action alternative and plays these off against other values that are inherent in the situation. Then imagination asks: Are there possible, latent, or emerging conflicts? Finally, dramatic rehearsals *highlight constraints* embedded in the action situation. In an exercise described in the section five below, students see how actions take place in a sequence where the earlier ones set the stage for and constrain later ones. Dramatic rehearsals develop and exploit sensitivity to how situations pose constraints that can limit or block the realization of even the best laid plans.

Mark Coeckelbergh has worked with dramatic rehearsals in the context of the Capabilities Approach. He examines how composing fictional scenarios would allow engineers to dramatically rehearse the consequences of adopting a particular technology (Coeckelbergh 2012). His example tests whether robotic technology can provide emotionally sensitive care to the elderly. Such scenarios give richness and content to deliberation so that in deliberation one both outlines the contours of an action and captures something of its emotional texture. Dramatic rehearsals cannot replace experience nor can they completely supplement a lack of pertinent experience in planning and executing engineering actions. But they extend experience through projections that are not unlike metaphorical projections; engineers can use experience (either what they have acquired themselves or have learned from others such as mentors) to structure and compose new, possible experiences. Dramatically rehearsing an action helps one prepare for its actual instantiation by giving one a sense of involvement in a situation as well as an existential feel for how the situation often pushes back against one's attempts to realize actions and plans.

Moral imagination should be integrated into engineering curricula as educators help their students prepare for the challenges posed by globalization. Through multiple framing, engineers achieve transperspectivity and break through prejudicial mind sets to *address other cultures from suitable points of view*. In order to adapt to *minds differently situated*, engineers role-take with others by projecting into their shoes and collecting their feel of a situation. Through dramatic rehearsals, engineers can test the implementation of decisions and designs in rich and emotionally textured worlds that are constructs of imagination; this helps develop and refine "that delicate knowledge of the heart." Moral imagination supports, builds upon, and even extends engineering expertise and experience to provide a response to global challenges that is socio-technically sensitive, interdisciplinary, emotionally responsive, and successful at integrating technical and moral expertise. When properly honed and deployed with other curricular components (and not deployed as a standalone), it provides an effective antidote to Esther's cautionary guidelines.

14.4 Morally Imaginative Engineering Agency

Moral imagination also contributes to teaching moral expertise in engineering. Huff, in a study of moral expertise in the computing field, has identified two modes of moral expertise, craftspersons and reformers. *Craftspersons* "tended to focus on their clients or users and to draw on pre-existing values in computing (e.g. user focus, customer need, software quality)....Thus, they tended to view themselves as a provider of a service or project...and to view difficulties or disagreements as problems to be solved" (Huff and Barnard 2009, p. 50). *Reformers* "tended to be crusaders who were attempting to change the values in social systems (organizations, professions, national cultures). They tended to view individuals as victims of injustice and to attempt to remedy that injustice" (Huff and Barnard 2009, p. 50). Huff presents both as viable, moral career tracks. What is important pedagogically is to direct students toward the mode of moral expertise that best fits their personality profile (Huff and Barnard 2009, p. 50).

Craftspersons and reformers can be locally or globally focused. Thus, an engineering curriculum addressing moral imagination should show how it contributes to (1) Locally-Focused Engineering Craftspersons, (2) Locally-Focused Engineering Reformers, (3) Globally-Focused Craftspersons, and (4) Globally-Focused Reformers.

Locally-Oriented Craftspersons work to provide a service to the client. They exercise imagination when identifying client interests and squaring these with the interest sets of the public, profession, and other engineers. Role-taking helps engineers identify a client's interests and distinguish *apparent* from *genuine* interests. It calls on empathy (the ability to see a situation through the client's eyes) while avoiding "egocentric projection" and "excessive appropriation" (Sherman 1998, p. 110). Transperspectivity comes into play as engineers question default mindsets that automatically frame situations and constrict thought and action; this skill helps engineers play over situations with different framings to unlock new possibilities. Deploying imagination properly empowers the engineer to say "no" to client directives when deeper understandings so dictate. Moral imagination exercised at this level of agency prevents the engineer from degenerating into a gun for hire who uncritically does the client's bidding. It also helps the engineer disagree with the client without breaking faith and to integrate client interests with those of other stakeholders.

Locally-Oriented Reformers also focus on the surrounding STS. But their agency targets the public who are vulnerable to engineering practice and the risks it entails.[8] The public is mostly unaware of the potential harms that accompany engineering practice; they are also not in a position to afford their consent. Locally oriented reformers step forward to identify the consequences of engineering projects and advocate for public wellbeing. Moral imagination aids carrying out this form of agency through role-taking and multiple framing. Role-taking with the public while armed with engineering expertise enables identifying public interests, latent and possible risks, and injustices caused by engineering projects and practice. Multiple framing makes it possible to question mindsets that cover over instantiated injustice and unacknowledged risk.

---

[8]Michael Davis defines the public as "those persons whose lack of information, technical knowledge, or time for deliberation renders them more or less vulnerable to the powers an engineer wields on behalf of his client or employer" (Davis 1997, p. 16).

Globally-Focused Craftspersons preserve values, provide services, and strive to bring about ethically sound results at the global level. Moral imagination is crucial because it provides ethical engineers with effective means for overcoming the pitfalls of telescopic philanthropy and clues engineers into the salient particulars of remote communities.

A case study shows how engineering expertise can be misdirected through lack of moral imagination. MIT engineers have designed a laptop computer (the XO model) for use in developing countries. They argue that laptop computers are ideal vehicles for delivering digitalized educational materials to children and that laptops can replace traditional teachers who work with old fashioned educational media like printed textbooks (Kraemer et al., p. 68). Yet, the XO laptop models disseminated through the One Laptop Per Child initiative have brought about mixed results. Education officials in the governments of developing countries remain unconvinced that laptops represent the best educational investment available; they balk at the high costs of teacher training and technical support (Kraemer et al. 2009, p. 70). Teachers in developing nations have resisted the introduction of laptops into their classrooms insisting that self-directed, computer-driven learning does not automatically trump community-based, low technology learning (Kullman and Lee 2012, pp. 45–47).

Those who have studied the introduction of technology into the classroom have already documented some of these problems (Cuban 2003). Viewing the adoption of a technical artifact critically through moral imagination helps to avoid such problems. XO laptop proponents could have developed fictional scenarios to imagine their implementation in different classroom settings. These scenarios would have anchored dramatic rehearsals for implementing laptops that would have uncovered some negative possibilities (Coeckelbergh). Overly restrictive mindsets may have blinded proponents to the downside of bringing laptops (and the self-directed learning pedagogy they embody) to teachers committed to community-based pedagogies. Role playing would have helped laptop proponents anticipate user discontent. For example, children in Ethiopia spend more time "playing and chatting" with their laptops than using them for learning. Students also rejected the software for composing music because it made no allowances for compositions that accord with local music traditions (Kullman and Lee 2012, p. 50).

While there is no question that the OLPC initiative is well-intentioned and backed by excellent engineering, there is a case to be made here that more strenuous efforts at exercising moral imagination could have anticipated and deflated many of the concerns outlined above.

Globally-Focused Reformers can first direct their efforts at redressing or repairing past injustices, especially those brought about by ill-conceived engineering projects. Reform efforts can also target preventing injustice stemming from inappropriate technologies proposed by well-meaning, but imaginatively deficient, outsiders. Paternalism is always a pitfall in these situations especially when there is a temptation to impose an advanced technology on a community not ready for it; in such cases, an appropriate technology (or what Schumacher calls an intermediate technology) would be called for. Of help here is the Capability Approach which can guide efforts at role-taking, multiple framing, and dramatic rehearsals (Nussbaum

2011). The Capability Approach asks would-be reformers to stop viewing globally remote communities as deficient or needy. Instead, engineers should reframe these communities as valuable repositories of knowledge and skill embodied in traditional practices and attitudes. The Capability Approach encourages professionals to view their projects and technologies as "conversion factors" that transform general capabilities into concrete functionings (Robeyns 2005, p. 99). How effectively they do this given the personal and environmental constraints in a given globally remote context provides crucial information on how appropriate the project or technology is (Oosterlaken et al. 2012, pp. 119–122).

Roopali Phadke provides an example in her report on how retired engineers, working with NGOs in India, were able to break a 14-year stalemate between the Indian government and the residents of several small villages in the rural province of Maharashtra (Phadke 2005). The government had proposed a large scale irrigation project centered on a dam that would flood the land occupied by four small villages and surrounding subsistence farms. The government and its staff of civil engineers regretted this sacrifice but claimed it was offset by the greater good. In response, local residents organized and successfully halted construction of the dam and reservoir.

NGO engineers broke through this stalemate by playing the role of honest broker. First, they worked with sociologists to teach local village residents participatory resource mapping to outline current land uses and to identify community values and capabilities. Thus, even though they did not carry out all the different disciplinary roles themselves, they formed effective alliances with sociologists and local villagers to perform this essential socio-technical analysis. Next, "they worked "to translate the data that emerged from a local mapping project into a technical alternative that complied with the agency norms…." (Phadke 2005, p. 508). Engineers do not have to be jacks of all trades to work effectively in interdisciplinary alliances. They do have to display communication skills, leadership capabilities, and team work practices that make it possible to work with others who have complimentary talents. Working effectively in such alliances, engineers can broker large scale projects to realize local values, preserve traditional ways of life, and integrate local concerns with the broader aims typical of large scale engineering projects. Moral imagination combines with each of these skills to compliment it, give it depth, and provide effective direction.

14.5 Teaching Moral Imagination

Teaching moral imagination does not require a radical departure from methods traditionally used to teach practical and professional ethics. But it does require redeployment of these in situations that allow for practice and responsive feedback (Frey 2010, pp. 621–624). The following activities have been used to enable students to practice moral imagination skills. They deploy formal and informal writing, verbal communication, role-playing, and structured reflection.

14.5.1 Teaching Responsibility Through "Dramatic Rehearsals"

The Hughes Aircraft Case module provides multiple avenues for exploring issues in engineering and computer ethics, including issues in peace engineering and appropriate technology (Huff and Frey 2008). In this exercise, students carry out a dramatic rehearsal to test different forms of dissent (Dewey 1960, p. 135).[9] Six turning points taken from the case show how the decisions made at each have placed events on a clear trajectory toward whistle-blowing. Students are invited to revisit these and dramatically explore different possibilities. What if participants had done X instead of Y? Would a different decision at a given point have led to different outcomes for the case? Students, in small groups, develop their alternative futures as "What if" dramas and act them out in front of the class. The decision points occur at different times and are dramatically rehearsed in chronological order. Arranging the "What if" dramas in this order constructs a dramatic narrative for the case that does not represent the case as it actually unfolded in the past but as it could have unfolded in a different possible world. This activity provides an opportunity for students to practice and reflect on moral imagination, especially dramatic rehearsals. But the reader must be cautioned against assuming that it creates this skill out of thin air or that it is a standalone activity that does not need future activities that help students to build further competence in this skill.

Like any skill-oriented pedagogy, dramatic rehearsals work best in building moral imagination if they are accompanied by structured exercises in reflection. Two such exercises bring this activity to a close. First, students reflect on their dramatization and create a storyboard. Some groups, after acting out their drama before class, act it out again and take photographs of key scenes. They then put this together, add dialogue, and arrange the scenes to build a narrative. Other groups draw their storyboards by hand to portray their dramatization as they remembered it. Either way, students are prompted to reflect on the narrative background of their drama and to rehearse concretely elements like character, action, conflict, background, fulfillment (or disappointment), and constraints.

Students also prepare written responses to questions that encourage them to reflect on their dramatic rehearsals. One question has them explore their drama as an experimental test of different forms of responsible dissent. In the actual case, the participants chose whistle-blowing. But what if they had tried to implement another form of dissent? Another question has students reflect on the form taken in their dramatic rehearsal. Was it a tragedy, comedy, documentary, cautionary tale, silent movie, or Quixotic adventure? (These are some of the forms students have enacted in the past.) Students are also asked to reflect on the conflicts that were played out in their dramas. If they were portrayed as value conflicts, were they

---

[9]Fessmire (2003) provides a useful summary of Dewey's dramatic rehearsals in *John Dewey & Moral Imagination: Pragmatism in Ethics*, pp. 69–91.

able to fully or even partially integrate the different sets of values? Was compromise necessary or even possible? The last question has students identify and reflect on the constraints they had to deal with in their dramatizations. For example, later decision points, because they were shaped in part by prior decisions, were more constrained. Students were asked to reflect on constraints and to analyze the strategies they used to work around them.

When students imaginatively play through alternate possible worlds in the form of "What if" dramas, they begin to see that the case could have had very different results if different decisions had been taken at key points. They also see the importance of dramatic rehearsal in deliberation. Dramatizing events gives students an opportunity to practice and reflect on role-taking, multiple framing, and, in this way, builds on their existing skills in exercising moral imagination.[10]

14.5.2   *Alternative Career Job Fair*

As part of the GREAT IDEA project (NSF 1033028), investigators designed and taught a course in Appropriate Technology.[11] They evaluated the ten modules used in this course in terms of seven objectives, the last of which was, *"Be able to competently undertake employment or research in appropriate technology."* While instructors consciously targeted this objective in the teaching of the course, students ranked it lowest in 9 out of 10 modules and in the overall course average (Castro et al. 2012). There are undoubtedly many reasons for this gap. But consider the hypothesis that students did not see the modules as offering alternative careers because they were under the influence of a mindset that framed a career as a well-paying job working for a corporation. To test this hypothesis, an alternative job fair was carried out in the fall of 2012.

The activity began by having students fill out a questionnaire that probed their openness to alternative careers in engineering and business:

- They were asked about the triple bottom line that adds environmental and social concerns to profits. Did they believe that companies could develop sustainable business models that integrated environmental sustainability and social justice with long-term financial stability?
- Students ranked salary, opportunity for promotion, support for continuing education, work environment, ethics policies, and environmental goals in terms their importance when considering where to choose to work. In general, the

---

[10]The author, in a recent publication, has addressed issues involved in teaching value traits, especially concerns to make abstract value statements more concrete through challenges and teamwork. See Frey and Cruz (2013).

[11]Appropriate Technology: Towards Sustainable Wellbeing, INTD 5095. This course has been taught in the second semester of academic years 2011–12 and 2012–13 by Marcel Castro and Chris Papadopoulos at the University of Puerto Rico at Mayagüez.

questionnaire probed how open they were to the idea that a good career involved more than a good salary.

Most students participating in the alternative career activity had also participated in the university's traditional job fair. The questionnaire asked them if this was a positive or negative experience. Then they were asked to weigh a satisfactory career against salary. Which did they find most important in their job and career search?

Students also participated in a video conference with representatives from two NGOs involved in relief work and community development. Each gave a short presentation followed by a question and answer session with student and faculty participants. Representatives also outlined the skills and knowledge required by community development projects and appropriate technology:

- Those working in relief efforts and community service need to develop professional and occupational skills and work from a deep commitment to helping others.
- Problem-solving skills are critically important, especially the ability to specify open-ended problems that arise in ill-defined situations.
- Conflict mediation skills were also called for since relief work and community development require finding common interests among differing stakeholders.
- Presenters emphasized the importance of being able to learn from past experiences, especially from failures. This learning should focus on particulars unique to the targeted socio-technical system.

One other skill set can be added to this list. Patricia Werhane et al. argue that building alliances between communities, governments and for-profit organizations is essential to tackling problems in poverty and community development; each organizational form has strengths that complement weaknesses in the others. Building these alliances requires proficiency in *moral imagination* as well as *systems thinking* and *deep dialogue* (Werhane et al. 2010).

More data is needed to confirm the hypothesis that students fail to see alternative careers because of overly restrictive mental models. But this activity provides opportunities to practice moral imagination. Students experiment with different ways of framing a viable career, questioning habitual mindsets that place priority on salary and prestige. Role-taking and techno-socio sensitivity are important in identifying opportunities for community and appropriate technology projects; this helps students identify non-traditional career opportunities based on community service. What is particularly powerful about this activity is that students are highly motivated to work through career possibilities; this provides an excellent opportunity to guide this process through the exercise of moral imagination.

14.5.3 Service Learning

In service learning, students provide a service to a community and learn skills and knowledge related to their area of academic specialty. Correlating service with learning is absolutely essential (Pritchard 2000).[12] Equally important, students should learn to employ frameworks (ethical, value-realizing, problem-solving, socio-technical) that help them recover and appropriate more fully what they have learned in performing community service.

GREAT IDEA offers graduate students service learning opportunities with the community of Duchity, Haiti on projects such as assessing a micro hydro-electric project, modifying an existing generator operating at 15 % capacity, and designing and implementing a new generator; all of these represent options to improve upon the status quo of 3 hours of electricity per day. Other projects include providing education modules to an orphanage on engineering and mathematics, rebuilding water infrastructure, designing and distributing dry toilets, and providing local residents with sand filters to clean water. Two courses have been developed in association with these activities. *Appropriate Technology* has been discussed by project PIs (Castro et al. 2012). A second course, *Responsible Research in Appropriate Technology*, will provide students with opportunities to reflect upon and further develop the skills they have practiced in community service projects.

The course, *Responsible Research in Appropriate Technology*, also raises issues in research ethics such as paternalism, informed consent (see especially Schrag 2006 for collective informed consent), and working to realize a community's capabilities (Nussbaum 2011). Moral imagination plays a central part in the design and execution of this course.

- The first module, "The Gray World," introduces graduate students to RCR issues (Rios et al. 2013). Students practice moral imagination through case discussion and role-plays. *Layered cases* present students with a straightforward, black-and-white core scenario. Then ambiguity is "layered" onto the core scenario by adding complicating circumstances. This type of case encourages multiple framings. Also employed are what Patricia Werhane terms *"Rashomon" cases* (Werhane 1999, pp. 69–88). Rather than providing a central, dominant narrative, these cases break down into a series of distinct and often inconsistent

---

[12] In *Needs and Necessities*, Baillie et al. (2010) talk about the "decoupling of the learning potential of service learning projects from the project itself as well as the potential harm they may cause within communities...." (Baillie et al. 2010, p. 105) From the context, it appears that they are arguing, as is asserted in this paper, that service and learning must be coupled in the community development projects designed to teach global engineering practices. But they also argue that most service learning projects benefit students more than the targeted developing communities (Baillie et al. 2010, p. 104). Their injunction that service learning must help students to learn "to think critically and historically" fits in nicely with the general argument of this paper. Critical thinking emerges from multiple framing. Historical thinking, according to philosopher of history, R.G. Collingwood, involves the use of "historical imagination" in the "reenactment" of past thought (Collingwood 1946/1974, pp. 231–248).

participatory perspectives. Rashomon cases elicit a careful review of and comparison between different participatory perspectives in which students exercise their skills of role-taking and multiple framing.

- In the second part of the course, students will work through a Technology Choice module. Studying different cases in technology choice (the OLPC case and India Irrigation project case described above have been used in past instantiations), students describe their case's STS, examine the technologies involved in terms of the Capabilities Approach, and teach their case to the class through a poster presentation.
- Finally, students will turn to developing case studies of their own in community service and appropriate technology. They will use a module on case study writing that discusses different uses of cases, how to identify good case topics, and how to carry out a comprehensive ethical and socio-technical analysis of the case.

Responsible Research in Appropriate Technology, will use these projects as the occasion for imaginative and reflective exercises designed to appropriate service learning in its concreteness and complexity. Moral imagination is not its only objective, but doing research in appropriate technology requires developing and refining skills such as role-taking, multiple framing, and dramatic rehearsals.

14.6 Conclusion

Moral imagination can be taught. Integrating it into a curriculum that extends engineering ethics to the global realm does not require a revolutionary redesign of existing classroom strategies. Rather it requires adopting a skills-based pedagogy where students practice skills of moral imagination in a context which encourages constructive feedback and deep reflection.

Role-taking, multiple framing, and dramatic rehearsal are the skill sets targeted in a pedagogy of moral imagination. These respond directly to pitfalls in global engineering, pitfalls foretold by Charles Dickens as his character Esther hesitates to embark on global crusades because of the challenges of adapting her mind to those differently situated, learning to address other cultures from suitable points of view, developing that delicate knowledge of the heart, and learning not to confide in good intentions alone. Moral imagination addresses these challenges to form an essential part of moral expertise as exercised in engineering in both local and global contexts.

Case studies, a mainstay in traditional ethics education, can be modified to serve as the occasion for instruction in moral imagination. Large cases can be broken down into decision points which students can explore through contrary-to-fact, "What if" dramas. Simple, core scenarios can be made more complex by layering in complicating circumstances which allow students to practice multiple framing. Cases where the core narrative is broken down into distinct participatory perspectives allow students to practice role-taking as well as multiple framing. In short, as one examines the teaching activities portrayed above, one gets a sense of how to deploy existing teaching materials and methods to directly address the challenges of teaching moral imagination.

There are limits to what can be brought about by moral imagination. First, it may not be possible to completely adopt oneself to minds that are differently situated; in these cases it may be necessary for an engineer to work as a member of an interdisciplinary team that includes cultural anthropologists, religious practitioners, sociologists, public policy experts, and, most importantly, local brokers. Second, those trained in Western traditions may have trouble addressing other cultures from suitable points of view. (GREAT IDEA investigators have found it difficult to assess energy-generating technologies from a suitable point of view; villagers may not have enough experience to dramatically rehearse what they would do with 24/7 electrical service.) And disparities between different cultures and locations may place severe limits on the ability to empathize and feel with others whose emotions and sensitivities have been partially shaped by different surrounding socio-technical systems. Moral imagination has not been put forth as a magic bullet solution to problems that arise through globalization. But this is not an all-or-nothing proposition. Imagination can and should be put alongside other moral objectives in formulating a curriculum in engineering ethics for globally oriented engineers. Finding the right combination is absolutely important but also feasible.

Acknowledgements The author gratefully acknowledges the support of the National Science Foundation under Grant No. 1033028 (NSF/EESE). He would also like to thank project PI, Christopher Papadopoulos and Co-PI Marcel Castro for their vision in designing the GREAT IDEA initiative and for the work they have done in Haiti to implement its ideas.

References

Baillie, C., Feinblatt, E., Thamaw, T., & Berrington, E. (2010). *Needs and feasibility: A guide for engineers in community projects – The case for waste for life*. San Rafael: Morgan & Claypool.

Bilbao, G., Guibert, J. M., & Fuertes, J. (2006). *Ética Para Ingenieros*. Sevilla: Desclée De Brouwer.

Bilger, B. (2009). Hearth surgery: The quest for a stove that can save the world. *New Yorker*, Dec. 21, 2009, p. 88.

Callahan, D. (1980). Goals in the teaching of ethics. In D. Callahan & S. Bok (Eds.), *Ethics teaching in higher education* (pp. 61–80). New York: Plenum Press.

Castro, M., Papadopoulos, C., Frey, W., & Huyke, H. (2012). Sustainable wellbeing education in engineering. *2012 IEEE international symposium on sustainable systems and technology*. DOI 10. 1109/ISSST.2012.6228005, 16–18 May 2012.

Coeckelbergh, M. (2012). How I learned to love the robot: Capabilities, information technologies, and elderly care. In I. Oosterlaken, & J. van den Hoven (Eds.) *The capability approach, technology and design* (pp. 77–86). New York: Springer.

Collingwood, R. G. (1946/1974). The historical imagination. In *The idea of history* (pp. 231–248). Oxford, UK: Oxford University Press.

Costello, A. (2010). Troubled water. *Frontline World*. July 29, 2010. http://www.pbs.org/frontlineworld/stories/southernafrica904v/video_index.html. Accessed 23 Apr 2013.

Cuban, L. (2003). *Oversold and underused: Computers in the classroom*. Cambridge, MA: Harvard University Press.

Davis, M. (1997). *Thinking like an engineer: Studies in the ethics of a profession* (pp. 16 & 55–59). Oxford: Oxford University Press.

Dewey, J. (1960). *Theory of a moral life* (p. 135). New York: Holt Rinehart.

Dickens, C. (1852/1985). Bleak house. New York: Bantam Books.

Fessmire, S. (2003). *John Dewey & Moral imagination: Pragmatism in ethics* (pp. 69–91). Bloomington: Indiana University Press.

Frey, W. (2010). Teaching virtue: Pedagogical implications of moral psychology. *Science and Engineering Ethics, 16*(3), 611–628.

Frey, W., & Cruz, J. (2013). Value integration: From educational computer games to academic communities. *Technology and Society Magazine, IEEE, 32*(1), 31–35.

Frey, W., & O'Neill-Carillo, E. (2008). Engineering ethics in Puerto Rico: Issues and narratives. *Science and Engineering Ethics, 14*(3), 417–431.

Frey, W., Papadopoulos, C., Castro-Sitiriche, M., Zevallos, F., & Echevarria, D. (2012). *On integrating appropriate technology responsive to community capabilities: A case study from Haiti.* Proceedings: American Society for Engineering Education, 2012: AC 2012–5106.

Gardner, J. (1978). *On moral fiction.* New York: Basic Books.

Harris, C. (2008). The good engineer: Giving virtue its due in engineering ethics. *Science and Engineering Ethics, 14*(2), 159.

Huff, C. (2001). Why a socio-technical system. *ComputingCases.org.* http://computingcases.org/general_tools/sia/socio_tech_system.html. Accessed 23 Apr 2013.

Huff, C., & Barnard, L. (2009). Good computing: Moral exemplars in the computing profession. *IEEE Technology and Society Magazine, 28*(3), 47–54.

Huff, C., & Frey, W. (2008). The Hughes Whistleblowing case. In R. Reena (Ed.), *Whistleblowing: Perspectives and experiences* (pp. 75–80). Nagarjuna Hills: Icfai University Press.

Huff, C., Barnard, L., & Frey, W. (2008). Good computing: a pedagogically focused model of virtue in the practice of computing (parts 1 & 2). *Journal of Information, Communication and Ethics in Society, 6*(3 & 4), 246–316.

James, W. (1890/1950). *The principles of psychology.* New York: Dover.

Johnson, M. (1986). *The body in the mind: The bodily basis of meaning, imagination, and reason.* Chicago: University of Chicago Press.

Johnson, M. (1994). *Moral imagination: Implications of cognitive science for ethics.* Chicago: University of Chicago Press.

Kraemer, K., Dedrick, J., & Sharma, P. (2009). One laptop per child: Vision vs reality. *Communications of the ACM, 52*(6), 66–73.

Kullman, K., & Lee, N. (2012). Liberation from/Liberation within: Examining One Laptop per Child with Amaryta Sen and Bruno Latour. In I. Oosterlaken & J. van den Hoven (Eds.), *The capability approach, technology, and design* (pp. 39–56). New York: Springer.

Nussbaum, M. (2001). *Upheavals of thought: The intelligence of emotions* (pp. 304–327). Cambridge, UK: Cambridge University Press.

Nussbaum, M. (2011). *Creating capabilities: The human development approach* (pp. 20–33). Cambridge, MA: Belknap Press of Harvard University Press.

Oosterlaken, I., Grimshaw, D., & Janssen, P. (2012). Marrying the capability approach, appropriate technology and STS: The case of podcasting devices in Zimbabwe. In I. Oosterlaken & J. van den Hoven (Eds.), *The capability approach, technology and design* (pp. 113–134). New York: Springer.

Phadke, R. (2005). People's science in action: The politics of protest and knowledge brokering in India. In D. Johnson & J. Wetmore (Eds.), *Technology and society: Building our sociotechnical future* (pp. 499–513). Cambridge, MA: MIT Press. 2009.

Pritchard, M. (1996). *Reasonable children: Moral education and moral learning.* Lawrence: University of Kansas Press.

Pritchard, M. (2000). Service-learning and engineering ethics. *Science and Engineering Ethics, 6*(3), 413–422.

Rios, C., Frey, W., Jarrimilla, E., & Echeverry, M. (2013). Accepting the likehood of ambiguity and disagreement on moral matters: Transitioning into the grey world. *Teaching Ethics, 13*(2), 55–71.

Robeyns, I. (2005). The capability approach: A theoretical survey. *Journal of Human Development, 6*(1), 93–114.

Schrag, B. (2006). Research with groups: Group rights, group consent, and collaborative research commentary on protecting the Navajo people through tribal regulation of research. *Science and Engineering Ethics, 12*(3), 511–521.

Sherman, N. (1997). *Making a necessity of virtue.* Cambridge, UK: Cambridge University Press.

Sherman, N. (1998). Empathy and imagination. *Philosophy of Emotions, Midwest Studies in Philosophy, 22,* 85–96, 110.

Smith, S. (2005). *Ending global poverty.* Palgrave Macmillan.

Urban-Walker, M. (2006). *Moral repair: Reconstructing moral relations after wrongdoing.* Cambridge, UK: Cambridge University Press.

Werhane, P. (1999). *Moral imagination and management decision-making.* Oxford, UK: Oxford University Press.

Werhane, P., Kelley, S., Hartmen, L., & Moberg, D. (2010). *Alleviating poverty through profitable partnerships: Globalization, markets and economic well-being.* New York: Routledge.

Winter, S. (1990). *Bull Durham* and the uses of theory. *Stanford Law Review, 42,* 639–693.

Chapter 15
Sifting, Winnowing, and Scaffolding: Structured Exploration for Engineering in a Modern World

Sarah K.A. Pfatteicher

> *Whatever may be the limitations which trammel inquiry elsewhere, we believe that the great state university of Wisconsin should ever encourage that continual and fearless sifting and winnowing by which alone the truth can be found.*
> University of Wisconsin Board of Regents, 1894

Abstract Over a century ago, in response to a complaint about a faculty member's teaching strategies, the University of Wisconsin Board of Regents issued a statement declaring that the purpose of higher education was to foster unfettered exploration with the faith that both faculty and students, if given the time and tools, would decide for themselves what to believe. The Regents' now-famous phrase "sifting and winnowing" emphasized the importance of the process of discovery, rather than mastery of rote knowledge, as crucial to true education, and served as the foundation for academic freedom as we know it today. The educational notion of sifting and winnowing, born in a time of clashing political philosophies and cultures, can serve as a guide for building engineering ethics curricula for a newly globalized world. But this exploration need not mean a free-for-all. Scaffolded learning can bring structure to the process of building students' capacity for independent, creative work in a wide variety of circumstances and cultures. This combination of organization and autonomy, counterintuitive though it may seem, provides an ideal grounding for globalized engineering practice.

Keywords Academic freedom • Critical thinking • Cultures • Ethics • History • Scaffolded learning

S.K.A. Pfatteicher (✉)
University of Wisconsin-Madison, Madison, WI, USA
e-mail: sarah.pfatteicher@wisc.edu

© Springer International Publishing Switzerland 2015
C. Murphy et al. (eds.), *Engineering Ethics for a Globalized World*, Philosophy of Engineering and Technology 22, DOI 10.1007/978-3-319-18260-5_15

15.1 Sifting and Winnowing

On July 5, 1894, Wisconsin State Superintendent of Public Instruction Oliver E. Wells wrote a letter to his fellow members of the University of Wisconsin Board of Regents complaining that University of Wisconsin economics professor Richard T. Ely had "taught socialism and other vicious theories to students at the University" (Herfurth 1948). After considering the testimony presented before them, the full Board of Regents boldly responded that the very purpose of higher education was to foster unfettered exploration with the faith that both faculty and students, if given the time and tools, would sort out the good from the bad:

> Whatever may be the limitations which trammel inquiry elsewhere we believe the great state University of Wisconsin should ever encourage that continual and fearless sifting and winnowing by which alone the truth can be found. (Herfurth 1948)

One hundred and twenty years later, one might take the Regents' declaration as a quaint commentary on an outdated form of higher education, no longer suited to the drastically altered circumstances of the modern world (Editorial 2015). In 1894, fewer than half of all practicing American engineers held a college degree. Not a single American engineering professional society had yet adopted a code of ethics. The significant engineering accomplishments of the day included the Brooklyn Bridge in 1883 and the first all-steel framed skyscraper (all of 10 stories tall) in 1889. The world's first subway (in London) had only existed since 1863; the patent on Alexander Graham Bell's telephone had been issued in 1876; Thomas Edison's first large-scale electric power station came online in 1882. The transportation and communication technology of the time meant that engineering and, indeed, business more broadly was a predominantly localized activity.

The world has changed dramatically since 1894. Today, even high school students can readily collaborate with their peers the world over. Transatlantic travel today is faster than trans-state travel a century ago. Multinational corporations are now commonplace, as are intercontinental engineering teams, bringing together diverse cultures, standards, and expectations in our daily work. Our society is far more globally connected than ever before, and the pace of innovation does not allow for lingering. One might suppose that in this newly interconnected and complex world, surely the task of educating our students has become more complex as well. We must prepare them for interactions and activities that would have been unimaginable even a generation ago.

But in many ways, Ely's world was changing as dramatically as is ours today. The world war that would erupt two decades after the Ely case resulted in part from shifting political and economic forces already emerging in the late nineteenth century. Disagreements and protests do not tend to emerge in a culture that is in steady state, but instead bubble up from disorder and uncertainty, often from a desire by a subset of the population to hold on to a comfortable, familiar past. Superintendent Wells' complaint against Professor Ely was not simply a clash of two strong personalities; rather, it was a sign of growing tensions between an old and familiar order and a new set of perceptions: between capitalism and socialism;

between West and East; between a provincial education aimed at traditional Wisconsin boys headed for traditional Wisconsin jobs and a global education for a brave new world (about which Aldous Huxley would soon write).

Read in this light, the Regents' decision exonerating Professor Ely is not only an elegant defense of the tradition of academic freedom, but also a reminder of our unchanging academic mission – to encourage students and society not to recoil from what is new or foreign to us, but to examine it carefully and choose our opinions and actions with purpose and reflection (Evans 2015; Hansen 1998). Indeed, "sifting and winnowing" is merely a more modern version of Aristotle's assertion that "it is the mark of an educated mind to be able to entertain a thought without accepting it."

A more recent dispute illustrates the ongoing need for such reminders. In 2006, University of Wisconsin provost Patrick Farrell received a complaint about a history lecturer who had been hired to teach a course on Islam and who was accused of claiming that the 9/11 attacks had been the result of U.S. government actions, not terrorism. In clearing Kevin Barrett to continue to teach, Provost Farrell recalled the Ely case, noting:

> Our students are not blank slates. They are capable of exercising good judgment, critical analysis and speaking their minds.... Instructors do not hand over knowledge wrapped up in neat packages. Knowledge grows from challenging ideas in a setting that encourages dialogue and disagreement. That's what builds the kind of sophisticated, critical thinking we expect from our graduates. (Provost 2006)

Sifting and winnowing remains a sensible educational philosophy because at heart it argues that education is less about transferring knowledge and more about transferring a set of skills by which knowledge can continually be gained, updated, and adapted. As the regents observed in 1894,

> We cannot for a moment believe that knowledge has reached its final goal, or that the present condition of society is perfect. We must therefore welcome from our teachers such discussions as shall suggest the means and prepare the way by which knowledge may be extended, present evils be removed and others prevented. We feel that we would be unworthy of the position we hold if we did not believe in progress in all departments of knowledge. (Herfurth 1948)

In short, the regents understood that an idea's currency is not the best measure of its accuracy. They argued that to censure Professor Ely's approach "would be equivalent to saying that no professor should teach anything which is not accepted by everybody as true," which the board refused to consider, as it would "cut our curriculum down to very small proportions." (Herfurth 1948) In 1957, at a rededication ceremony for the Sifting and Winnowing plaque commemorating the Regents' report, Helen C. White, professor of English, spoke anew of knowledge not as a static thing to be grasped and held, but as an ideal toward which to strive:

> In the free give-and take of the University students get a vision of what a lifelong undertaking the pursuit of [truth and wisdom] is. Indeed, I think that that is the most valuable thing we give them on this campus. For there is only one thing more important than the preservation of freedom, and that is its use. ("Sifting and Winnowing," 2006)

In the early twenty-first century, the "lifelong pursuit of truth" so valued by Professor White in 1957 remains at the core of higher education, even for fields far removed from economics and English. At the college level, no English literature class would ask students merely to memorize the text of a novel without exploring and analyzing its meaning and power in the culture of the time. No history course would require students to memorize dates of key events without also asking them to consider why those events matter and to analyze what led to them happening when and where they did. In a similar vein, although engineering surely rests on foundational knowledge of the language of mathematics and the laws of physics, these basics only take on meaning when they are applied in a specific context and for a specific purpose. As civil engineering students quickly learn, no two bridges are ever exactly alike, and it is in the differences that engineers' professional qualifications come to matter. Equations of force may remain constant, but our understanding of soil mechanics and wind loads evolve, as do the size and weight of the vehicles that will travel on our spans, and the materials with which we can build them.

It has become a truism in engineering that much of what undergraduate engineering students learn in school will be out of date within 5 years of their graduation. Engineering faculty and administrators frequently bemoan the challenges of packing ever more content into the limits of a 4- or 5-year curriculum. But these views are founded on the assumption that the primary purpose of an engineering degree is to fill students' brains with a full complement of the knowledge they will need for their careers. How different a view of engineering education is suggested by the 1894 Wisconsin Regents – that the goal is to teach the skills required for the *pursuit* of knowledge, not the mere capture of it. If students leave an engineering program knowing how to sift and winnow to identify what they need to know at any given time and for any given project, we have given them far more than if we simply gauge their factual mastery at the moment of their graduation.

The goals of a modern, global education in engineering ethics should not be so very different from our goals for engineering education more broadly. There are codes and standards that guide the practice of both engineering and engineering ethics, and yet these vary with place and time and circumstance (Pfatteicher 2003). Consider, for example, changes over the past century in American engineers' acceptance of bribery as a business practice (American Society of Civil Engineers 2007). When it was first adopted in 1914, the code of ethics of the American Society of Civil Engineers did not address bribery at all. The ASCE code revisions approved in 1963 included a canon known as the "when in Rome" clause that permitted engineers to adapt their practice of bidding for contracts to the country in which they were operating. By 1977, the Foreign Corrupt Practices Act forbade such relativism, and in anticipation of that law, ASCE had removed the "when in Rome" clause in 1976. Canon 6 now declares that civil engineers "shall act with zero tolerance for bribery, fraud, and corruption." Comparably dramatic changes have taken place in building codes, labor practices, environmental standards, and liability law (see, for examples, Wermiel 2003; Garcia 2009; Weiss 2012; White 1996). Our students should know that what is considered acceptable now may be frowned upon or even illegal by the time they are in mid-career, and what is normal in their home context may be out of bounds in their work locale.

The codes we teach students today may not be the ones they will need tomorrow. The question is: how do we teach them enough about such standards to develop their capacity to apply them appropriately and also to be able to locate the relevant codes for any future situation? Indeed, how do we teach them to respect and value current standards of practice while also encouraging them to view such standards with a critical eye? The goal for our students, whether in engineering design or engineering ethics, should be to hone their talent for gauging when a given code or standard or approach is appropriate, reliable, defensible, and when it should be viewed with caution or even skepticism. Call it sifting and winnowing, call it lifelong learning, call it critical thinking. Whatever term we use, it is this ability to judge, gauge, and assess – in short, to filter rather than merely accumulate – that should be the primary purpose of undergraduate education in general and of engineering ethics in particular.

Educating students for a globalized world adds to our efforts the challenge that we prepare students for a whole world of opportunities, not for narrow and predictable career paths. My students might be studying in Madison, but I must prepare them for practice wherever they land, whether that be Milwaukee, Moscow, or Mumbai. (Indeed, a fair number of my students in Madison now come from overseas.) The form of sifting and winnowing students must be able to practice if their work carries them to different countries and cultures presents even greater challenges. It is not enough for them to understand, for instance, that bribery is considered unacceptable by current U.S. standards if they are working in a location where payment for favors is considered *de rigueur*. They may find themselves working for a company like Google, balancing the desire to grow their customer base in China with a reluctance to facilitate government censorship of Internet searches (Jacobs and Helft 2010). Or they may be overseeing manufacturing for a company like Apple or Walmart, balancing the goal of producing affordable products with the right of factory workers to a living wage and humane working conditions (Stiglitz 2007). Or they may be working for a construction company in Turkey or Japan helping rebuild after a devastating earthquake, and needing to abide by local building codes for resilience and redundancy that may be weaker or stronger than those taught in California (Tang 2000, p. 3; Glanz and Onishi 2011). They need not only to understand in the abstract the potential conflicts that exist between cultures' expectations, but also to determine in practice how to resolve or cope with such conflicts. And they must do so in a way that allows them to remain true to their own values.

I contend then that the practice of ethical engineering requires a well-honed ability to sift and winnow, to investigate, question, and decide for oneself which actions are better and worse in any given set of circumstances, not simply the capacity to memorize codes or standards to be followed mindlessly. Accreditation standards in the U.S. require that engineering programs must demonstrate that their graduates have "an understanding of professional and ethical responsibility" (ABET 2013). As I have argued elsewhere, "strictly speaking, the criterion does not require programs to demonstrate that graduates *are* ethical; it requires that they *understand* professional and ethical responsibilities." (Pfatteicher 2001, 2005) In that context, I noted that our goal should be to teach, rather than preach, ethics to our students, and that we should assess their knowledge and skills, not their behavior *per se*, for the solutions they design may not be the ones we envision for them.

If we thought the purpose of engineering ethics education was to ensure our students' mastery of all the codes and standards they would need, like so many equations into which the specifics of a case study could be plugged, we would indeed face a daunting instructional task. But as the 1894 regents so perceptively grasped, our purpose in higher education should not be to demand that our students master a given set of facts. True, our examples may change with time, but that indeed is all the more reason not to test students on the specifics of the examples, but on the process used to think through those examples. The ever-changing nature of society and of engineering practice is itself the very reason not to focus our teaching on such specifics (Pfatteicher 2010).

Skilled sifting and winnowing may be our goal, but how are we as educators to train ourselves and our students in the habits of mind that will serve well in bad times as well as good, in international as well as domestic situations, in today's circumstances as well as tomorrow's?

15.2 Scaffolded Learning

A half-century ago, cognitive psychologist Jerome Brunner first spoke of scaffolding theory, arguing that children's language acquisition was fostered by parents providing decreasing levels of structured support as the child's skill increased. (See, for example, Bruner 1961.) In the decades since then, other scholars built on the concept and applied it to a broad array of formal educational settings, developing theories of contingent, embedded, reciprocal, and technical scaffolding, to name but a few variations. The important point for our present purposes is that scaffolded learning has long been encouraged as a means to help students master challenging concepts and skills.

My own introduction to the concept of scaffolding came not from the literature, but from a desperate parental need. I had a 2-year-old son who, like all 2-year-olds, was fond of the phrase "I do it myself!" And yet, like toddlers everywhere, his desire to do whatever the "it" of the moment was – from buckling the latch in his car seat to brushing his teeth – was often at odds with his ability to complete the task unassisted.

We turned to books on child development and parenting in search of help. Learning that this dilemma was productively taxing his brain on a daily basis was comforting to us as parents. That determined drive to learn to do it ourselves is what helps us through the often painful – or at least uncomfortable – process of learning new skills. Realizing that this internal battle was in fact the foundation of his important lifelong development helped too. But understanding our toddler on an intellectual level was of little practical use at 7 a.m. as we struggled to get him dressed and out the door to daycare. Certainly, if we gave him all day, he would eventually get some clothes on, though we had no assurance they would be the right ones for the weather. But we did not have all day. And trying to forcibly dress a toddler is like trying to put pantyhose on a panicky octopus.

What eventually saved us was a parenting book tip that we came to call the Red Shorts/Blue Shorts method. Simple, really. Rather than opening the closet and asking our son what he wanted to wear (and rather than just choosing for him, which would only lead to a battle of wills), provide a limited set of choices, any of which is acceptable, and let him choose. Do you want the red shorts or the blue shorts today? The choice remained his, but within bounds that made it a manageable decision. And one day our son surprised us by asking "Can I wear my green shorts instead?" He had learned the model of what was OK and wanted to adapt it. Success!

I do not mean to suggest that undergraduates are childlike, only that novice learners of all ages benefit from expert-provided structure. (See, for example, Lipscomb et al. 2004). As a National Research Council report noted,

> What children can do with the assistance of others is even more indicative of their mental development than what they can do alone [because] what a child can perform today with assistance she will be able to perform tomorrow independently, thus preparing her for entry into a new and more demanding collaboration. (Bransford et al. 2000, p. 81)

In that sense, my toddler is little different from undergraduate engineering students. If they have not been exposed to a task before, how are they to know the consequences to consider? Just as my son lacked the experience to grasp the abstract concept of temperature, students new to ethics can hardly be faulted for their lack of familiarity with contractual obligations, liability law, or disciplinary precedent, all of which may affect one's judgment of how to proceed when faced with a choice. And engineering ethics is at its core about choices.

Professional ethics concerns itself with those scenarios individuals face because of their specialized knowledge and skills, particularly when those skills are applied on behalf of or for the benefit of those who do not possess that specialized knowledge. Elementary ethics may set out basic standards and rules: loyalty to an employer is good; falsifying data is bad. While important, these concepts hardly challenge the typical engineering undergraduate. Dilemmas in professional ethics arise when a situation pits these basic tenets against each other: what should you do when your supervisor asks you to alter the records of testing results? This scenario is not simply about right versus wrong; it presents a tension between two competing values: integrity of the data and integrity of the relationship. Nor is this simply worded case black-and-white; there are many possible reactions one might propose, not all of which are defensible, but several of which may be considered "ethical," and some subset of those that may be more practical, effective, or realistic than others.

Education research has shown that a scaffolded approach helps students develop independent, expert skills. How might we provide enough scaffolding to support students as they explore such options, without oversimplifying the challenge before them? In the National Research Council report cited earlier, John Bransford and his colleagues catalogued significant differences between experts and novices, noting, "experts have acquired extensive knowledge that affects what they notice and how they organize, represent, and interpret information from their environment. This, in turn, affects their ability to remember, reason, and solve problems." The goal of scaffolded learning, whatever form it takes, is first to "engage learners … by focusing their attention on critical elements." (Bransford et al. 2000, pp. 31 & 68.) Well-constructed scaffolding helps focus attention on the key details, temporarily masking distractions from view. (In this sense, a more apt metaphor might be the blinders worn by racehorses, but scaffolding is the term of art.)

Scaffolded learning builds on the notion that our nature as humans is to crave structure and we can be paralyzed by too many choices (Schwartz 2004; Scheibehenne et al. 2010). And yet an excess of structure can deaden our ability to make choices, or lead to rebellion against the very structure we are trying to impart. After all, 18–20-year-olds are themselves navigating a more advanced version of the "I do it myself" stage so common to toddlers. So how might we apply the concept of scaffolding to engineering ethics education? What is the right balance of structure and flexibility in our curricula to foster the habits of mind appropriate to ethical practice of engineering education in a globalized world?

The scaffolding looks different at 20 than at 2, as it should. And the ultimate goal, of course, is to remove the scaffolding altogether, revealing an adult skilled at the decision-making and judgment required in daily life. If the purpose of engineering ethics instruction is not to require mastery of a given code or set of expectations, but to foster an ability to employ principles, examples, and skills to think critically and creatively through any dilemma, we must build those expectations into our curricula.

Let us turn, then, to consider two counterintuitive aspects of combining scaffolding with sifting and winnowing. First, one might assume that providing structure for students would limit their creativity, a talent critical to engineering work. Second, one might assume that providing structure would artificially ensure success for students, protecting them from the valuable lessons to be gained from the occasional failure. Surprisingly, perhaps, research suggests that both of these assumptions are faulty.

15.2.1 Does Scaffolding Limit Creativity?

Engineers have long taught a structured process of design. Indeed it is this engineering problem solving that has been touted as the most valuable skill engineers have to offer fields outside their own: the ability to take a complicated problem, break it down into its component parts, simplify each one by making some considered assumptions, deal with each piece separately, then re-combine them into a more complex whole.

Building on this strength, engineering ethicists in the 1990s recommended adapting this structured problem-solving technique to the analysis and solution of ethical dilemmas. Philosopher Caroline Whitbeck first advocated for this parallel:

> The multiply constrained nature of many problems in engineering design provides an excellent model of challenging ethical problems involving many types of moral considerations, all of which must be taken into account. Many ethical problems that are represented as conflicts are better understood as problems with multiple demands and ethical constraints, constraints that may or may not turn out to be simultaneously satisfiable. (Whitbeck 1998)

One of the notable strengths of the ethics-as-design concept is that it highlights the idea that "solving" ethical dilemmas (like solving design problems) is less about finding the one right solution and more about creating **a** workable solution that meets certain criteria. The best solutions rely on a student's ability to imagine multiple options and then sort through them to select and refine one – to sift and winnow ideas and to choose from among them.

Building on Whitbeck's work, engineer Charles Fleddermann employed this ethics-as-design model in his introductory textbook on engineering ethics:

> Ethical problem solving shares … attributes with engineering design. Although there will be no unique correct solution to most of the problems we will examine, there will be a range of solutions that are clearly right, some of which are better than others. There will also be a range of solutions that are clearly wrong. There are other similarities between engineering ethics and engineering design. Both apply a large body of knowledge to the solution of a problem, and both involve the use of analytical skills. So, although the nature of the solutions to the problems in ethics will be different from those in most engineering classes, approaches to the problems and the ultimate solution will be very similar to those in engineering practice. (Fleddermann 2012)

Philosopher Lisa Newton took the steps of the design process and, using the letters of the word DISORDER, translated them into an 8-step process for reasoning through ethical dilemmas (Newton 1998):

- Dilemma – *define it.*
- Information – *acquire it.*
- Stakeholders – *identify them.*
- Options – *explore them.*
- Rights/Rules/Results – *consider them.*
- Decision – *make one.*
- Effects – *evaluate them.*
- Review & Reconsider – *to evaluate the outcomes and to learn from them.*

Students who have been exposed to the engineering design process will find the steps here familiar: define the problem, document the customer's needs, develop potential solutions, compare the options, and so on. As in design, the benefit of such steps is to provide a clear and structured path to follow when faced with a complex and ambiguous problem. And as in design, novice learners will lean more heavily and consciously on such steps than expert practitioners, who follow the same process, but automatically and unconsciously (Whitbeck 1998; Bransford et al. 2000).

In a scenario such as the employer's request to falsify data, then, scaffolding might come in the form of questions to ask of the situation. Should loyalty ever outweigh honesty? How important is safety? Does the culture permit a subordinate to question an order? How serious are the consequences? Do you have sufficient experience and context to know why your supervisor is asking this? How exactly should you phrase whatever response you choose? These questions require higher-order skills in Bloom's Taxonomy: analysis, synthesis, evaluation. The steps of the DISORDER mnemonic provide a structure and a sequence to such questions, helping students move from "these are the rules" to "what would you do?"

The value of this device is that it builds on a skill set that engineering students are already developing in their technical coursework, and that it can be applied to a wide range of ethical dilemmas:

- How do you balance designing to serve immediate local needs with designing to protect against potential global effects? (And how prescient should you be expected to be?)
- If you accepted a job offer with a company in a country other than the one in which you were raised, how would you balance your personal ethics with the professional expectations of the company? (Indeed, how do you learn what those expectations are?)
- Under what conditions would you feel compelled to design for a level of safety that goes beyond what is required by local standards? (And how do you respond when your supervisor has set those design limits for you?)[1]

A 2008 study of the effects of this scaffolding methodology showed that the structure of the DISORDER device enhanced students' ability to think creatively about approaches to ethical problem solving, and that the effects did not seem to be affected significantly by the length of the presentation. The solutions students identified at the beginning of the module, before they had been exposed to DISORDER, were limited in number, varied in practicality, and were occasionally even illegal. The solutions students created with the help of the DISORDER mnemonic were greater in number and more likely to be both viable and legal. Though one might imagine the imposed structure would limit the creativity of students' thinking, instead it freed the students to work on the desired skill set by keeping them from becoming frustrated by the vastness and vagueness of the problem. As scaffolding theory suggested, the external structure supported their ability to screen out irrelevant information and wrestle with the core issues (Masters and Pfatteicher 2008).

---

[1]Catalogs of engineering ethics cases addressing these and other issues are available from the Online Ethics Center (onlineethics.org), the National Institute for Engineering Ethics (niee.org), and the American Society of Civil Engineers (www.asce.org/Ethics/A-Question-of-Ethics/), among others.

15.2.2 Does Scaffolding Artificially Ensure Success?

The engineering profession paradoxically embraces failure even as it seeks to ensure success. Henry Petroski, for one, has argued that "to engineer is human" – as humans, engineers will make mistakes and we would be foolish to expect otherwise (Petroski 1985). But that need not mean that we accept or excuse such failure. Petroski is hardly alone in pushing for engineers to study failures with determination, lest we repeat the deadly errors of the past. Can we provide scaffolding for students without artificially eliminating failure from the learning process?

In 1990, the American Society for Engineering Education (ASEE) issued a challenge to engineering schools to "develop or expand . . . first-year entry programs [that] introduce students . . . to the spectrum of opportunities in engineering and provide them with engineering experiences." (Hoit et al. 1998) Few American universities at the time offered such courses in the first year; many assumed that without adequate knowledge of calculus and physics, students did not have the skills to tackle a complex design problem, the archetypal engineering experience. And yet these same schools faced the challenge of encouraging students to stick with the rigorous foundational work long enough to discover the rewards of more real-world engineering work later in the curriculum.

The ASEE challenge spurred a variety of curricular experiments. Engineering schools responded in the decade that followed with a wide array of introductory courses, many of which included a hands-on introduction to engineering design (Smith 1993). By their nature, such courses require that we trust our students, letting them explore failure and learn from it. It is through this process of trial and error that they begin to hone their ability to judge and assess solutions. As the ASEE committee surmised, we give students a markedly different vision of engineering education and engineering practice when we begin with open-ended challenges rather than the drill of countless problem sets whose certain answers lie in the back of the book. The challenge, though, was to provide enough support for novice engineers to engage in design productively without over-constraining their exploration.

In the mid-1990s, a group of engineering faculty at the University of Wisconsin-Madison was among those who took up the ASEE challenge and developed a new design course for freshmen engineering students. The group of UW-Madison faculty had recently completed a 9-month team-based professional development program called "Creating a Collaborative Learning Environment" where they studied, discussed, and reflected on the learning process. This program was designed and facilitated by a human factors engineer with expertise in fostering organizational change through faculty development. At the conclusion of the development program, the college administration asked this faculty group to design a course to improve the retention of freshmen engineering students, with specific emphasis on the retention of underrepresented student groups. As their facilitator later explained,

> The faculty decided to place students in groups to satisfy real client requests from the surrounding community. Students would then be asked to design, build and present their project to the client, so they could have a complete experience, connecting modeling, design and calculations to the unexpected and varied issues that come into play in the construction stages of design. The faculty hoped that by serving a client with real needs, many times requiring customized design (e.g., the design of an adjustable, folding desktop for use on a wheelchair, a guide dog harness for a person with multiple sclerosis), students would see that what they were doing had meaning to others outside the university. If they accomplished their design task well, it might positively affect someone's day-to-day life. They might see a direct link between engineering's content and design process to the wellbeing and welfare of people in the surrounding community. Engineering might become, for some students, more expressly relevant to society—and to themselves. (Sanders et al. 2006)

The course is now 20 years old and has become a core part of a required first-year curriculum for engineering undergraduates. Although many details have changed, the course continues to emphasize providing students with the information and experience necessary to make informed decisions about whether engineering is the right field for them. Students still conduct real-world design projects with community clients, and are still expected to begin to develop "the habits of mind that engineering study and practice require."

Given a full semester, students have the luxury to work through an entire project from conception to building. An ongoing challenge in directing the course is to remind the laboratory instructors that failure is an acceptable part of the experience for students. Successful scaffolding requires that we be willing to let go as instructors, recognizing that our support is intended to be temporary – left in place only until the students are able to stand on their own. Our competitive and perfectionist tendency is to want our students' design to "win" – to work well, to impress the client, to earn praise from the other instructors. But just as my son needed to learn for himself the consequences of not taking a sweatshirt on a 50° day, design students need the room to make poor choices and struggle to recover from them. Just as in ethics, some actions are not okay. Entering the tool shop without safety gear and proper footwear? Not acceptable. Incorrectly calculating the strength needed for the support structure and having to re-calculate the figures and re-build the final product? Perfectly acceptable. Mistakes are okay; injuries are not. The teams that tend to thrive are those where instructors set high expectations, step in briefly when needed, and encourage students to keep at it when they struggle.

In the end, this approach allows students to demonstrate what they have learned, not to regurgitate what they have been told. Instructors make resources available (texts, manuals, experts chosen by us as relevant), but students are encouraged to seek out more or different resources as they need to. The sifting and winnowing is theirs to do.

The success of this model is most evident in what happens to students after the course is over. A major goal of the UW course (like many such courses) had been to improve the retention of engineering students, particularly traditionally underrepresented students. Within a few years after the course began, retention from freshman to sophomore year for students taking the course was 5 % greater than for a control group of students who did not. For women and minority students, the increase in retention approached 10 %. And not only were students who took the course retained, their subsequent grade point averages were roughly half a point higher than their peers in the control group who had not taken the course. Graduation rates for students who began their undergraduate careers in this course also outpace their peers who did not have this introduction to design. Little surprise, then, that the course remains one of the most popular in the engineering curriculum, among students, faculty, and administrators. (Little surprise, too, perhaps, that one of the original faculty designers of the course was none other than Patrick Farrell, later provost of the university during the Kevin Barrett case described above.)

What this success masks, however, is that students are not graded on the success of their projects per se, but on their engagement with the problem and their ability to work through the inevitable challenges that arise in real-world design. Above all, the course presents engineering as a complex and ambiguous undertaking with no single "right" answer (in contrast with many of their first-year courses, where problem sets still reign supreme). Students' grades are based on their ownership of the process as much as on the success of their design. To earn a high grade, students are better off designing something themselves, even if it fails, than borrowing a design from elsewhere, no matter how successful. Note that few of the projects have an explicit international component, and yet the independent thinking and self-confidence nurtured by the course support students' ability to operate in a wide variety of contexts. In addition, although students are not always explicitly asked to identify ethical dilemmas in their work, the fact that the projects require negotiation with a client and assessment of the tradeoffs of multiple design options means they are wrestling with issues of professional ethics whether or not they recognize it in the moment. If, in working together to explore and expand the edges of their abilities, students have grasped something of the challenges of true design, they have, as Helen C. White noted a half-century ago, benefitted from the "free give-and take" that characterizes the best of higher education.

15.3 A Vast Diversity of Views

The ideal educational environment is one where students feel safe in exploring but also uncomfortable enough that they must push themselves beyond their current abilities to succeed. This notion of growth through feeling both safe *and* uncomfortable is at the heart of learning and indeed is the basis for tenure, a system similarly intended to encourage exploration and creativity by providing security during the pursuit. It is the sifting and winnowing that the Board of Regents recognized and defended over a century ago.

At the time, the regents spoke of the presumed need to satisfy the citizens of the state that funded their university:

> As Regents of a university with over a hundred instructors supported by nearly two millions of people who hold a vast diversity of views regarding the great questions which at present agitate the human mind, we could not for a moment think of recommending the dismissal

or even the criticism of a teacher even if some of his opinions should, in some quarters, be regarded as visionary. (Herfurth 1948).

How much harder if instead of satisfying the "nearly two millions of people" in Wisconsin we believed we must meet the test of what every member of the globe holds to be true? What the regents understood in 1894 remains true 120 years later: we should not attempt to prepare students for specific locales or job functions or beliefs. Laws change. Codes evolve. Standards adapt. Customs vary. If we build it properly and well, our engineering education ought to be transportable to anywhere a student might land. Far from altering education to fit a globalized world, we must continue to take as our purpose to teach our students what will serve them well *anywhere* they find themselves, no matter what "vast diversity of views" they encounter along the way.

References

ABET. (2013). 2013-14 criteria for accrediting engineering programs. http://abet.org/accreditation-criteria-policies-documents/

American Society of Civil Engineers. (2007). A question of ethics: Payment of "consulting fees" to foreign agent to secure contracts. *ASCE News*.http://www.asce.org/Publications/ASCE-News/2007/10_October/A-Question-of-Ethics/. Accessed 25 Aug 2013.

Bransford, J. D., Brown, A. L., & Cocking, R. R. (2000). *How people learn: Brain, mind, experience, and school*. Washington, DC: National Academic Press. http://www.nap.edu/openbook.php?isbn=0309070368. Accessed 22 June 2013.

Bruner, J. S. (1961). The act of discovery. *Harvard Educational Review, 31*, 21–32.

Editorial Board. (2015). Gov. Walker's 'Drafting Error.' *New York Times*. 7 Feb A18.

Evans, Christine. (2015). Save the Wisconsin Idea. *New York Times*. 16 Feb A17.

Fleddermann, C. B. (2012). *Engineering ethics* (p. 5). Upper Saddle River: Pearson Prentice Hall.

Garcia, M. (2009). China's labor law evolution: Towards a new frontier. *ILSA Journal of International and Comparative Law, 16*, 235–253.

Glanz, J., & Onishi, N. (2011). Japan's strict building codes saved lives. *New York Times*. 11 March, A1.

Hansen, W. L. (1998). *Academic freedom on trial: 100 years of sifting and winnowing at the University of Wisconsin–Madison*. Madison: University of Wisconsin Press.

Herfurth, T. (1948). Sifting and winnowing. Resource document. State Historical Society of Wisconsin. http://digicoll.library.wisc.edu/WIReader/WER1035-Chpt1.html.

Hoit, M., & Ohland, M. (1998). The impact of a discipline-based introduction to engineering course on improving retention. January, 79–85.

Jacobs, A., & Helft, M. (2010). Google may end venture in China over censorship. *New York Times*. January 13, A1.

Lipscomb, L., Swanson, J., & West, A. (2004). Scaffolding. In M. Orey (Ed.), *Emerging perspectives on learning, teaching, and technology*. http://projects.coe.uga.edu/epltt/. Accessed 22 June 2013.

Masters, K. S., & Pfatteicher, S. K. A. (2008). *Lowering the barriers to achieve ethics across the engineering curriculum*. Proceedings of the American Society for Engineering Education Annual Conference.

Newton, L. (1998). Doing good and avoiding evil. http://www.rit.edu/cla/ethics/resources/manuals/dgae1p6.html. Accessed 30 June 2013.

Petroski, H. (1985). *To engineer is human: The role of failure in successful design*. New York: St. Martin's Press.

Pfatteicher, S. K. A. (2001). Teaching vs. preaching: EC2000 and the engineering ethics dilemma. *Journal of Engineering Education, 90*(1), 137–142.

Pfatteicher, S. K. A. (2003). Depending on character: The ASCE shapes its first code of ethics. *Journal of Professional Issues in Engineering Education and Practice, 129*(1), 21–31.

Pfatteicher, S. K. A. (2005). Anticipating engineering's ethical challenges in 2020. *Technology and Society Magazine, 24*(4), 4–43.

Pfatteicher, S. K. A. (2010). *Lessons amid the rubble: An introduction to post-disaster engineering and ethics*. Baltimore: Johns Hopkins University Press.

Provost review clears Barrett to teach class on Islam. (2006). http://www.news.wisc.edu/12701. Accessed 30 June 2013.

Sanders, K., Farrell, P. V., & Pfatteicher, S. K. A. (2006). *Curriculum innovation using job design theory*. Proceedings of the human factors & ergonomics society 50th annual meeting.

Scheibehenne, B., Greifeneder, R., & Todd, P. M. (2010). Can there ever be too many options? A meta-analytic review of choice overload. *Journal of Consumer Research, 37*, 409–425.

Schwartz, B. (2004). *Paradox of choice*. New York: Harper Perennial.

Sifting and winnowing. (2006). *Letters and Science Today, 12*(1), 1–2. https://kb.wisc.edu/images/group86/21911/v12no1p1-2.pdf. Accessed 22 June 2013.

Smith, K. A. (1993). Designing a first year engineering course. In M. E. Schlesinger, & D. E. Mikkola (Eds.), *Design education in metallurgical and materials engineering: Engineering design in courses and curricula*. Warrendale, PA: The Minerals, Metals & Materials Society.

Stiglitz, J. E. (2007). Multinational corporations: Balancing rights and responsibilities. *Proceedings of the Annual Meeting: American Society of International Law, 101*, 3–60.

Tang, A. K (Ed.). (2000). Ismit (Kocaeli), Turkey, Earthquake of August 17, 1999, (ASCE) Technical Council on Lifeline Earthquake Engineering, Monograph #17, March 2000.

Weiss, E. B. (2012). International environmental law: An introduction. http://untreaty.un.org/cod/avl/ls/weiss_el.html#. Accessed 25 Aug 2013.

Wermiel, S. E. (2003). No exit: The rise and demise of the outside fire escape. *Technology and Culture, 44*, 258–284.

Whitbeck, C. (1998). Ethics as design: Doing justice to ethical problems. In *Ethics in engineering practice and research*. Cambridge: Cambridge University Press.

White, R. J. (1996). Top 10 in torts: Evolution in the common law. *Trial, 32*(7), 50–53.

Chapter 16
Toward a Global Engineering Curriculum

Eugene Moriarty

Abstract This paper distinguishes between Globalism as *theory* and Globalization as *practice*. I accept the existence and importance of the globalization process but aim to critique the resulting structure from a theoretical Globalism point of view. The specific theory I will emphasize is Critical Theory which arises from the Frankfurt School of philosophical deliberation. The question becomes how can the Critical Theory of globalism help to bring the liberation of globalization to all of humankind? The role of engineering in the rise of global structures is key to my concerns. Different kinds of engineering yield different ways of being global. I distinguish between Standard Engineering and Focal Engineering and the different kinds of engineering ethics arising from each of these. Then I conclude by offering a possible Global Engineering curriculum, unlike the traditional EE, ME, CE, etc. curricula, which addresses issues arising from a globalized world, issues like global warming and fossil fuel dependency. Such a curriculum will keep a solid first 2 years that will be similar to existing engineering curricula but incorporating a small subset of global ideas. Then the upper-division will be where the differences dominate, focusing on real global problems like world hunger and environmental destruction. The upper-division will be steeped in ethics, an engineering ethics program that introduces process ethics (the ethics of the process, which stems largely from Kantian Deontology and Utilitarianism), virtue ethics (the ethics of the person), and material ethics (the ethics of the product). A systems approach will also be emphasized in this curriculum. Finally Focal Engineering will provide the *ethos*, the characteristic spirit of the global engineering culture.

Keywords Globalism • Globalization • Virtue • Process • Product • Focal

E. Moriarty (✉)
San Jose State University, San Jose, CA, USA
e-mail: eugene.moriarty@sjsu.edu

© Springer International Publishing Switzerland 2015
C. Murphy et al. (eds.), *Engineering Ethics for a Globalized World*, Philosophy
of Engineering and Technology 22, DOI 10.1007/978-3-319-18260-5_16

16.1 Introduction

This paper comes to you from out of the realm of globalism which I take as the theoretical and conceptual side of the globalism/globalization constellation. Globalization is the actual practical happening of the event of the emergence of a global world, the coming into being – for better or for worse – of ever greater connections of economic, political, social, and cultural global dimensions. The theory side and the practice side of the globalism/globalization constellation complement each other. The theory reflects on the practice. Globalism reflects on the practices of globalization. Literary scholar Marshall Brown notes that there is some ambiguity in understanding globalism and globalization. Clear and distinct definitions prove to be somewhat elusive. The definitions above are a compilation of the views of several authors. Brown's proposed definitions are quite simple: "By globalism I understand an idea, an image, a potential; by globalization a process, a material phenomenon, a destiny." (Brown 2007, p. 143). Digging deeper: globalism is the thematization of the world "characterized by networks of connections that span multi-continental distances." (Nye 2002). Globalization as the practice side of the globalism/globalization constellation is that which is thematized by the theories and concepts of globalism. Another way to put this is that the theory of globalism is one's understanding of the global process, whereas the practices of globalization constitute the enactment of that understanding. The practices of globalization are often viewed as having positive and/or negative faces. Part of my aim in this essay is to incorporate into global engineering curricula a concern for ethical assessments of the processes of globalization, sorting out the positives and negatives, the pros and the cons.

A closer look at the globalism as theory and globalization as practice distinction reveals an intimacy between the two poles of the constellation. At the dawn of modernism, Francis Bacon advocated what came to be known as a marriage of theory and practice. Prior to Bacon's time there were purely theoretical ventures – like asking the question how many angels can dance on the head of a pin – and purely non-theoretical activities as well – like building a bridge using only intuitive understanding. According to historian of philosophy W.T. Jones, Bacon hoped to merge the strengths of theoretical conceptual understanding and practical intuitive understanding in the service of all humanity. His aim was to purge all knowledge, which he took as his province, "of two sorts of rovers, whereas the one with frivolous disputations, confutations, and verbosities, the other with blind experiments and auricular traditions and impostures, hath committed so many spoils, I hope I should bring in industrious observations, grounded conclusions, and profitable inventions and discoveries." (Jones 1969). The off-spring of that marriage led to numerous developments, for example, the disciplines of modern science and engineering. Engineering today is still a practical enterprise, as it was from ancient times, except that now its practices are grounded in theoretical discourses like mechanics, dynamics, thermodynamics, etc. The practice of engineering is grounded in theory but the theory is also grounded in the practice. The theory behind engineering sciences stems from practical considerations. For instance, engineering thermodynamics is both theory based and practice based, a very different affair than the purely theoretical physics version of thermodynamics. Modern engineering has always been a process which relies not only on the engineering sciences, but also on economics, typically focusing on design or other activities like testing, operations, maintenance, or marketing that serve the design process. The economics focus in engineering practice, always important in all aspects of engineering, is especially crucial in global processes like off-shoring. Off-shoring involves economics to a large degree, but much else as well. Politics and cultural differences are also necessary considerations. Engineering today is being more and more a globalized practice, a practice that seeks a theory, a theory that attempts to describe, to understand, and to prescribe positive paths for the practices of engineering.

In this essay I will look at the global nature of engineering and possible curricula that will bring that global nature to light and put it into practice under the auspices of a circumspective ethics outlook. Such an ethics outlook will seek to determine how the global engineer ought to be, how global engineering ought to be, and how globalized engineered products ought to be. Talk of the "ought" brings the is/ought distinction, fundamental to philosophic ruminations, into view. The "is" dimension is covered by the disciplines of physics, the social sciences, and metaphysics, and the "ought" dimension by ethics. Modeling global dynamics provides a description of the globalization phenomenon. From such models prescriptions arise. What can we do, what should we do, what must we do? For instance, the social dynamics of the US is connected to the social dynamics of China and India and a plethora of other places. These dynamics connect to the dynamics of global politics and global environmental concerns. Such models, if they cannot tell us exactly what to do, can point in the direction of what to do. Back in the 1970s, MIT Professor Jay Forester (1971) attempted to model the dynamics of the world in terms of feedback control systems, using just a handful of variables. The non-linearity inherent in most real systems, however, can add tremendous complexity to the entire process. How exactly global engineering and a global curriculum ought to be, in view of the difficulties with describing or modeling a dynamic and global world, is not clear. Still, we can develop linear models to provide points of departure for global systems models, approximations that stem from the contributions of the engineering and social sciences. But of particular concern will be to not equate a global engineering curriculum merely to its engineering science aspects which indeed do tend to be universal. An engineering science course, for example, in statics and dynamics taught in the US will be quite similar to statics and dynamics taught in China or Brazil. Rather, engineering must be seen as a global project which includes a wide range of engineering sciences but also "organizational, economic, environmental, social, and temporal elements, . . . those things that give engineering a reason to be, that define the problems to which it is directed, that dictate which elements of the scientific and instrumental content are appropriate to use, and that determine the pace and process of the engineering enterprise." (Newberry 2005, p. 9). Within the globalism/globalization constellation purview, universal and particular, as well as practical and conceptual, aspects of the engineering project contribute to a dynamic system of variables constituting a totality, a world dynamics. Perhaps a revival of Forrester's methodology is in order, especially in light of the fact that high order, non-linear, dynamic systems these days can be modeled without too much difficulty by simulation programs like MATLAB.

16.2 Globalization

To be sure, globalization is a major process impacting contemporary life. Under its sway, we recognize "the facts of the increasing interdependence of national economies, the diffusion of technology and technological activity across international borders, and the intersection and integration of cultures." (Newberry 2005, p. 9). Globalization is a process, a practice, an event, enlarging the domain of engineering activity, happening everywhere and sometimes all at once. As a *process*, globalization refers to the interaction and integration among the people, companies, and governments of different nations, a process driven by international trade and investment, aided by information technology. The *practice* of globalization is the execution of its processes. An *event* of globalization occurs when the practice produces a balance, a harmony, when all parties involved in a transaction express satisfaction. For engineering education to connect with globalization, students would need to be provided with "an enlarged set of knowledge and skills required to address the situations encountered in this larger domain. This includes inculcating foreign language skills, knowledge of foreign laws, practices, and customs, or knowledge of foreign environments, resources, and needs. (Newberry 2005, p. 10). Of course, we cannot expect a student of engineering to understand the depth of foreign law or the depth of a foreign language. How much of an exposure will suffice? How much knowledge needs to be accrued? Those become topics for discussion. In addition, explicit conceptual knowledge may not be enough. We may have to follow our intuitions, to take a stand, to say this is right, that is wrong. Providing reasonably salaried jobs for a 3rd world country is right, but leveling their forests is wrong. This is where the moral dimension becomes prominent. Globalization is directly involved because it aims at the good or ought to do so. As the engineers would have it, doing no wrong is one thing but actually seeking to do good is another. We will see that standard engineering seeks to do no harm, but what I am calling focal engineering seeks to bring good products into the world. How the good is interpreted is, of course, another story but a story that needs developing, that needs telling, if an ethical dimension is to have a prominent place in the global discourse. Globalism, the theory side of the globalism/globalization constellation, might be the framework within which the good can be discussed. In any event, the pros and the cons of the global venture need to be brought to light in authentic future global engineering curricula.

Douglas Kellner, UCLA Professor and critical theorist focusing on media theory, maintains that "a wide and diverse range of social theorists are arguing that today's world is organized by increasing globalization, which is strengthening the dominance of a world capitalist economic system, supplanting the primacy of the nation state by transnational corporations, and eroding local cultures and traditions through a global culture." (Kellner 1996, p. 1). This is surely the negative view of globalization. On the more positive side: "Globalization is the latest stage in a long accumulation of technological advance which has given human beings the ability to conduct their affairs across the world without reference to nationality, government authority, time of day or physical environment." (Langhorne 2001). Does globalization yield increasing homogeneity or does it produce heterogeneity through increased hybridization? There is a growing contingency that believes there is a dangerous, unsustainable, and unethical way that economic globalization is being accomplished. Corrupt government officials are giving away power to large corporations for their own personal gain: human rights/environment/health/economy/democracy all are at risk. Eternal and, indeed, global vigilance is becoming the price of liberty. Furthermore, Nigerian Political Science Professor Lucky O. Imade (2003) maintains that although globalization is certainly the buzzword of the new millennium, the nature and impact of globalization has been the subject of profound debate and concern in economic circles since the mid-1990s. Will unfettered global market forces increase or diminish the gap between the rich and the poor? Proponents of globalization insist it has promoted information exchange, led to a greater understanding of other cultures, raised living standards, increased purchasing power and allowed democracy to triumph over communism. For instance, on the negative side, echoing the above remarks of Douglas Kellner, Minnesota Secretary of State and agricultural activist Mark Ritchie (1996) says that globalization is merely the process of corporations moving their money, factories and products around the planet at ever more rapid rates of speed in search of cheaper labor and raw materials and governments willing to ignore or abandon consumer, labor and environmental protection laws. As a reality, it tends to be largely free of ethical or moral considerations.

Another way to look at globalization is in terms of the processes of exclusion and inclusion. Inclusion is often associated with the social, economic and political status quo. The inclusive power brokers employ these established systems to their own advantage. Excluded are the disenfranchised and that exclusion becomes ever more negative as globalization becomes ever more encompassing. As Scholte (2000, p. 53) argues, globalization has too often "perpetuated poverty, widened material inequalities, increased ecological degradation, sustained militarism, fragmented communities, marginalized subordinated groups, fed intolerance and deepened crises of democracy". Nevertheless, some globalization analysts seek to develop the positive aspects found within the same conditions of social and global order as those mentioned above. The question becomes how to foster greater human inclusion and emancipation. For example, the advance of technology which – among other things – drives the globalization process may create for the state sophisticated tools to spy on its citizens. But those same tools might be used for mobilizing and empowering those same citizens. The task, then, is to channel opportunities for inclusion that exist within globalization's wealth of contradictions. By sorting out options and possibilities, Globalism might be able to help with the empowering of a wide range of societal actors. Globalization benefits from scrutiny by Globalism.

16.3 Globalism

Though we may not be able to totally close the gap between positive and negative globalization, we might get some more clarity by moving from the globalization realm to the globalism realm. We might look at globalism as a complement to globalization. The suffix "ism" added to a word generally refers to the "idea about" or the "theory of." Globalism is a theory of globalization. The positive and negative effects of globalization can be examined from a theoretical perspective. A step back to a reflective state of mind results in an experience (*Erfahrung*) that lends perspective to globalization as it is immediately experienced (*Erlebnis*). Globalism provides the theory for the practice of globalization. But what kind of theory? The notion of theory can refer to simply an idea about the global world, or to something very complex like Marxist social theory, or more typically to something in between. A variety of theories contributes to various forms of globalism. There is structuralism, social constructionism, feminism, deconstructionism, postmodernism, hermeneutics, theories of the natural sciences, et al. The theory that appeals to me, because it is so wide ranging, and thus capable of enlivening the notion of globalism most efficaciously would be critical theory. It stems from what is called the Frankfurt School, which included thinkers like Marcuse, Adorno, Horkeimer, and Habermas. Within critical theory, as Marcuse (1973, p. 145) points out, "the real field of knowledge is not the given facts about things as they are, but the critical evaluation of them as a prelude to passing beyond their given form." But even the critical theorists, once they come up with an evaluation, still have the problem of how to turn their ideas into policy via political processes. All theories, it seems, must be taken with a grain of salt. The Frankfurt theorists studied "art, music, political economy, technology, the public sphere, and the rise of fascism." (El-Ojeili and Hayden 2006, p. 6). Though these concerns do not seem overtly connected, they do exhibit a subliminal common ground, and the thinkers of the Frankfurt School did possess a certain unity of purpose – namely, the attempt to move society towards rational institutions which would ensure a true, free and just life. (Held 2004, p. 15). All this is in spite of the fact that the critical theorists "share the view that the dominant discourses of modernity emerging from the Enlightenment social and political thought are in a state of crisis." (El-Ojeili and Hayden 2006, p. 9). The critical theorists aim in general for human emancipation, which means to seek the causes of, "and prescribe solutions to, domination, exploitation, and injustice." (Ibid., p. 10).

Critical theorists continue to seek a balance between their theory and their practices, between – for the sake of my interests in this paper – globalism and globalization. As Muqtedar Khan puts it: "faith and interest in globalism drives globalization – and, in turn, globalization spreads globalism." (Khan, 2003, p. 3).

Ideas drive the process and the process spreads the ideas. In addition, globalism provides a realm of discourse within which not only can the practices of globalization be described but optimal practices can be prescribed as pointing toward what we ought to do. The is/ought distinction emerges, an ethics emerges, an ethics that expands into the global arena. Mark Ritchie maintains, as mentioned earlier, the key aspect of globalization is that it is an economic process largely unfettered by ethical concerns. He views globalism, on the other hand, as the idea that we share a fragile planet in a reflective manner which requires mutual respect and careful treatment of all its inhabitants and its environment. Those are indeed ethical ideas. In addition, for the engineering project, globalism requires the daily active practice of a set of values and ethical beliefs emerging out of a consideration of what it means to be good, to do good, and to make good products. "Active communications to foster understanding, the sharing of resources on the basis of equity and sustainability, and mutual aid in times of need are three central activities that undergird globalism." (Ritchie 1996, p. 1).

The marriage of theory and practice that Bacon advocated can perhaps come to pass as globalization comes into balance with globalism. As practice needs theory so too does globalization need globalism. A final example from the realm of educational administration: assume administrators can function without a policy grounded in an explicit theoretical framework. They just react to whatever befalls them in an intuitive manner. Everything that happens is unique. There is no general framework of interpretation whereby particulars can be subsumed and shown to be of a certain general class and not another. The educational administrator who does not recognize the power of theory to guide us toward finding meaning in situations is at best a functionary of the overall educational system. One wonders how they might have been promoted to such high level jobs in the first place. Similarly, assuming that theory does not need practice is comparable to generating school policy without testing the policy by putting it (or parts of it) into practice. The policy ideas which spring from the theoretical structure of the educational system itself need to be "run up the flagpole." Would enactment of these policy measures generate an improved school atmosphere? Without a doubt, educational administrators need to develop solid theory based policy, but put into practice, attention needs to be given as to how the policy pans out.

16.4 Standard Engineering and Focal Engineering

A standard engineering course of study consists of 40–45 units of engineering sciences (thermodynamics, statics, computer programming, circuit theory, strength of materials, etc.), 40–45 units of General Education, and 40–45 units of specialized major work. Not much has changed in this general structure for well over 50 years. But recently, in view of the spread of globalization around the world, ABET (Accreditation Board for Engineering and Technology) has come up with some new requirements that point engineering curricula in a global direction. Both standard engineering and what I am calling focal engineering, in order to keep up with global transformations happening everywhere, can and should embrace global elements and incorporate them into their engineering curricula. As well as the more standard requirements, ABET now requires engineering programs to demonstrate that students can effectively serve on multi-disciplinary teams, communicate effectively, especially across international borders, and demonstrate an understanding of professional and ethical responsibility. Engineering programs must demonstrate that their graduates have the broad education necessary to understand the impact of engineering solutions in global, societal, environmental, and political contexts and they must demonstrate knowledge of contemporary issues within those contexts (ABET 2013). For example, even though an engineer may be an electrical engineer and not have detailed knowledge of the problem of global warming and its causes, he or she must have some sense of how global warming is affecting the planet and what engineers can do to contribute to the problem's solution. Globalization and globalism have both affected and continue to affect the standard and focal kinds of engineering practice.

Let us now take a look more specifically at what these structures of globalism and globalization have to do with the engineering project. Globalization as a practice has been around for many thousands of years. The same can be said for engineering. As the pre-modern era gave way to modernism – assume the shift into modernism occurred "around" 1500 AD – European adventurers, explorers, and colonizers began their global quests, which gave globalization a stimulus to thrive. Not only were Western products brought to the Americas and to the East, but also Eastern products were sent West. Gun powder, for instance, was a product of medieval China and was brought to the West and by the early modern era gunpowder was having a great influence on all aspects of warfare. Engineering contributed. Engineering thrived, especially a bit later in the modern era, from about 1780 to 1850. That was the golden era of modern engineering practice, the time of the Industrial Revolution. But today the modern era has started to give way to the postmodern era. Albert Borgmann (1993) indicates two major paths we can take as we move across the postmodern divide separating the modern era from the postmodern era. Although it is really impossible to pinpoint where and when we as a civilization have crossed, or are crossing, or will cross that divide, at least we can distinguish the pre-modern from the modern era and the modern from postmodern era. And what is the nature of the two possible paths characterizing the postmodern era? In particular, what is the nature of engineering and the nature of the global practices as we cross the postmodern divide? One path is just a continuation of business as usual, except at an intensified and accelerated level. Borgmann calls that hyper-modernism. This is more or less where we are today or toward where we are heading. The other path, more human and engaging, he calls the path of postmodern realism.

Several values characterize modernism, among them aggressiveness toward nature, the triumph of a universal method (Descartes'), and a stress on the individual. As we move on into the postmodern era, at least along the path of postmodern realism, these values shift toward more sensitivity to the environment, a greater concern for the particular rather than the general, and a new emphasis on community rather than on the individual. The shift in environmental attitude is probably most apparent, from Francis Bacon's view of nature as a phenomenon meant to be subdued for the benefit of humans, up to contemporary times when more and more of us see ourselves as part of nature, one with nature, and our engineered world needs to be resonant with that nature. Hence the rise in significance, in all engineering processes, of the value of environmental sustainability.

Engineering in the modern era I call standard engineering. As engineering crosses the postmodern divide, and as Borgmann would maintain, it becomes either hyper-modern engineering, which is standard modern engineering amplified and intensified, or a kind of engineering characterized by postmodern realism, which I call focal engineering (Moriarty 2008). Focal engineers engineer focal products which are products that are engaging, enlivening, and resonant. Consider, for example, the Golden Gate Bridge. I would call it a focal product because it certainly performs a valuable function in delivering the traveler from Marin County to San Francisco and back efficiently and safely. But as a focal product it does something else as well: it engages and enlivens the artist and the photographer, people are drawn to the Golden Gate, the bridge itself resonates with its locale. Of course there are many products that are not generally considered focal. They may be efficient and useful and good things to own. Consider, for example, the snow shovel that sits in my garage. I do not normally consider it focal though I am grateful to have it when I need it. But I am not drawn to my shovel, it is drawn to me, when I need to use it and call upon its services. Focal engineering asks not only "how" or "what" questions, like standard engineering, but also and especially "why" questions. It follows the Precautionary Principle: it abides by the rule that we should proceed with caution with regard to any new technology that is being proposed, for which we lack sufficient scientific and technical knowledge. For example, the engineering of GMO products should proceed with caution. Focal engineering seeks to bring forth products that exhibit what Borgmann calls a commanding reality instead of a disposable reality. It is concerned not just with the end-user and the product, but also with the world or the context out of which the product emerges and into which it coalesces. That world is the human lifeworld in which humans embrace a multiplicity of goals and enact a variety of roles and take up with all sorts of devices and things. Focal engineering relies crucially on conversations. By having honest, open, non-coercive conversations in the lifeworld, engineers, stakeholders, and all interested parties can hash it out and hopefully arrive at some kind of consensus.

Hyper-modern standard engineering is a form of postmodern engineering practice and has evolved in the past several decades into an inherently global phenomenon. For example, the globalization of hyper-modern standard engineering entails resources being gathered from Peru, designs being concocted in Germany, proto-types being constructed in Australia, testing being done in India, manufacturing being done in the Maquiladoras of Mexico. The resulting product is marketed across the globe. All of these activities strive for maximum efficiency. Globalization prevails in contemporary hyper-modern engineering practice.

Focal engineering is most often initially a local venture. It begins with people getting together in places like libraries, churches, hospitals, and meeting halls.

Conversations are paramount. Should we install another stop sign on this corner? What kind of shielding do these antennas need in order to keep our citizens safe? Should we ban nano-technologies from our community? If they are so small you cannot see them, how can they be controlled? What kind of engineered products do we need and choose to develop? Initially, we think locally and act locally. Reflection is fundamental to focal engineering. Globalism, not Globalization, is the home of the focal engineering venture.

Focal engineering is still primarily in the idea stage, in the realm of globalism, in the realm of possibility. One possibility emerging from the focal engineering reflection is the amplification of the empathic dimension in the focal engineering practice. What does it mean to be empathic? Empathic to the other, we walk a mile in their shoes. We look out for their welfare. We and they are on the same team. Lots of indications in recent years point to empathy as intrinsic to the human condition, the human psyche, the human biology. As one example, Jeremy Rifkin provides a wealth of information about how as a civilization we are shifting away from the overt willfulness that has characterized the modern era more toward an empathic civilization emerging in the postmodern era. (Rifkin 2009) Doing right by the other evokes ethics. What then might all this say about ethics, engineering ethics in particular?

16.5 Engineering Ethics

The engineering project is activated by engineers who bring requisite resources – materials, ideas, money, expertise – to bear on the tasks at hand, resulting in products of various kinds, products like bridges, networked communications, machinery, surveillance systems, etc. With the person of the engineer, the process of engineering, and the product, the engineered, I associate three different kinds of ethics and consider their significance within a global context and a global engineering program of study. The kinds of ethics involved are, first of all, Virtue Ethics, the ethics of the person, the engineer. Secondly, there is Process Ethics, the ethics of the process, engineering. Thirdly there is Material Ethics, the ethics of the product, the engineered. The most common type of engineering ethics is process ethics, sometimes called theoretical or conceptual ethics, stemming largely from Kantian deontological ethics and/or Utilitarianism. Often in engineering it is the only type of ethics employed for assessing the various processes of engineering practice. Virtue ethics, the ethics of the person, has been gaining popularity in the past few decades and is increasingly employed for the assessment of the character of the engineer. Material ethics, the ethics of the product of the process of engineering, stems from the work of Albert Borgmann and looks at whether the product is harmonious with its world and the users of the product. Process ethics and virtue ethics are necessary for the ethical assessment of engineering, but only with the incorporation of material ethics does the ethical assessment become sufficient.

The ethics of Standard Engineering is based largely on the admonition to "do no harm" to the world at large and the ethics of Focal Engineering is based largely on the admonition to "actually do good" with the products it brings forth into the world. The latter admonition includes the former. Standard Engineering practice follows the ethics of the person (Virtue Ethics) and the ethics of the process (Conceptual or Process Ethics). The ethics of the person and the ethics of the process are both typically encapsulated in Codes of Ethics. Focal Engineering in principle follows Virtue Ethics and Conceptual Ethics and, in addition, the ethics of the product (Material Ethics). How can we as engineers be morally good persons? How do we practice our engineering activities in a good way? And how do we make good products that will contribute to the engagement, enlivenment, and resonance of the end-user taking up with his or her products within his or her world?

The Standard Engineer engineers products that are safe and technically efficient. But are they good in an ethical sense? The idea seems to be that most products of the Standard Engineering enterprise are ethically neutral. A rifle, for example, is a fine hunting tool to help me feed my family. Or is it an evil machine for the slaughter of innocents? It depends on how the end-user uses it. Standard Engineering finds virtue ethics of the person and conceptual ethics of the process to be necessary. Environmental Sustainability, for instance, much considered as a value we need to stress these days, falls under the conceptual ethics of the process of engineering. This is good, this is necessary. But what's missing? As mentioned, I contend that we also need Material Ethics in order that our ethical reflections will become more complete, more sufficient. Material ethics assesses products or proposed products for how well they contribute to actually doing good in our now globalized and globalizing world. I have been suggesting that such products should be resonant, enlivening, and engaging. That is my sense of what constitutes the good. That's easy enough to say from within the comfortable realm of Globalism. But realistically how can we bring into the world products that will actually do good in the global world we inhabit? As a first step we can talk, especially about our cultural differences. Conversations get the ball rolling. We could then transcend, adopt a big-picture view, organize conferences that gather together stakeholders and all interested parties, follow perhaps a model like the Danish Consensus Conferences. But the more immediate problem is to bring these issues and concerns into the engineering curriculum so that students and instructors can get familiar with the language and procedures of global discourse.

16.6 Curricula

We are all familiar with the curricular constraints under which we labor. Many of us have probably tried to incorporate some ethics into our already overcrowded curricula. Just any concern with ethics is usually seen as a competition for academic turf. A little here and a little there, sure. But not a whole course. And then there is also the problem of the global nature of engineering today. What does that entail? How do we educate students to be cultured and globally civilized engineers? At Purdue University there is the opportunity to join global design teams which provide the opportunity for students and faculty to engage in international projects. The University of Rhode Island offers an International Engineering Program which consists of a BS in an engineering discipline and a BA in a language. This is a 5-year program. Baylor University incorporates globalization experience into their engineering programs by requiring two courses. The first course is a combination of a Tech writing course and an engineering economics course, and the second course focuses on technical entrepreneurship in a global economy.

Lots of other universities are providing global experiences via modified coursework and this is certainly a good thing. My preferences are along the lines of what Byron Newberry (2005) has to say. He suggests a more radical approach: create whole new programs. Let the standard BS Engineering programs remain, perhaps modified slightly to incorporate global concerns, for instance, via minimal ABET requirements. But for Global Engineering I suggest we focus on new Programs. Perhaps they can constitute a new department, the Department of Global Engineering. They might keep more or less the structure of the first 2 years but have freshman and sophomore engineering science courses and general education courses incorporate global concerns. A simple electrical engineering circuit design problem, for example might be to figure out a resistance value and a capacitance value that will yield a desired cutoff frequency for a low-pass filter. The problem might be prefaced by a short narrative that shows how low-pass filters are used in the generation of a switching mechanism that is used in a water filtering system for a third world country. The upper division, junior and senior level courses will be where the real differences lie. Global Engineering will be focused on the problems of world hunger, the lack of education, relieving the suffering of the impoverished of the world, overcoming the lack of control over one's own destiny, and solving the problem of the environmental destruction we see all around us. Meeting the needs of the developing world will require a very different upper division program than a program intended to help our students match the skill sets of other engineers in the global marketplace.

What would a Global Engineering upper division look like? It must be more practical and less theoretical than the typical engineering curricula. It would cover the engineering of basic technologies that can directly benefit developing countries. It must adopt an approach "that can better incorporate cultural, environmental, and ethical considerations." (Newberry 2005, p. 13). Global justice and sustainability must be part of that ethical concern. What is urgently needed, according to Hans Kochler (2000, p. 7), is a globalization of moral awareness "in regard to the basic rights of each individual and each nation, and of the obligations resulting from the mutual recognition of those rights."

A systems approach would be essential. Woven throughout the program would be courses in global economics, global philosophy, global politics, etc. History would also be involved, the history of the global movement especially. Along this line I would argue for the inclusion of some of the ideas of the great global thinkers of the past. Global thinkers are systems thinkers. Systems thinking flourished from the time of WWII to the present. It has been applied in numerous contexts. Now-a-days nobody makes a big deal of it. It needs a new challenge. I think the global movement presents that challenge.

For example, ought not global engineers consider the systems work of Jay W. Forester, mentioned earlier, especially his *World Dynamics*? Forester's model for World Dynamics, I believe, has been under-utilized. His model tracked five key variables and offered a set of differential equations whose solution represented the evolution over time of those variables. The variables Forester chose were population, food production, industrial production, pollution, and the consumption of natural resources. My suggestion is directed toward junior and senior global engineering students: each semester of upperdivision studies take the Forester model and extend it by two variables (e.g. the average level of sea rise or the average amount of methane in the atmosphere). Finding appropriate data for these new variables will be a task in its own right, but incorporating it into the Forester model should be a straightforward assignment. The senior project could involve comparing the original Forester model with the new enhanced model the student comes up with. Such activities and studies will also constitute a true interdisciplinary segment of the global engineering curriculum, because the engineering students will need to interact with environmental studies folks, mathematicians, social scientists, et al.

Other systems thinkers whose work might bear resurrection include Gregory Bateson, Norbert Wiener, Barry Commoner, Warren McCulloch, R. Buckminster Fuller, Claude Shannon, John von Neumann, W. Ross Ashby, Ervin Laszlo, Ludwig von Bertalanffy, Kenneth Boulding, C. West Churchman, Donella Meadows, Heinz von Foerster, Niklas Luhmann, James Lovelock, and a host of others. Barry Commoner, for instance, provided a real boost to the systems world with his famous statement that "everything is connected to everything else." As simple and true and eloquent as that statement is, today it has very little traction. The Cartesian methodology at the dawn of the modern era forced us to look at breaking things down, at disconnecting, at abstraction. Now at the end of the modern era perhaps the Cartesian methodology has run its course. As we enter the postmodern era perhaps the time has come for the holistic, connected, big picture view to begin to call for our attention.

The Global Engineering Department can be the home of Focal Engineering, aiming at the creation of focal products. These are products that directly benefit humanity, all of humanity, as we are all connected by threads and cables of flows and empathic forces. The focal product engages and enlivens and resonates with the world in which it is used. Or if the product is not focal in itself but serves a focal practice, then it also contributes to the benefit of humanity. For example, I may have a focal practice of connecting with others via social media for the sake of sharing information and strengthening empathic energy. I may use the internet to advance my focal practice. Then that usage contributes to the overall benefit to humanity. Focal products emerge from the focal practices of focal engineering, and Focal Engineering practices are practiced and studied within a Global Engineering curriculum.

16.7 Conclusions

Developing the Global Engineering curriculum will be a monumental task. Do we take it back to square one and start building it up from there? I suggest and have suggested in previous paragraphs a more modest approach:

1. Keep the first 2 years similar to what they are now, but factor global ideas in here and there. Require 2 years of a foreign language. Discuss foreign laws, practices, and customs. Here would be a good place to introduce the notion of a global code of ethics, currently under development, as in the work of Heinz Luegenbiehl. Discuss foreign environments, contexts, resources, and needs. For instance, in a standard problem in a thermodynamics course present the context as a global phenomenon which, say, points out some environmental concern.

2. The Global Engineering upper division program will be focused on the problems of world hunger, the lack of education, relieving the suffering of the impoverished of the world, overcoming the lack of control over one's own destiny, and solving the problem of the environmental destruction we see all around us.

3. Upper division courses must stress ways to better incorporate ethical considerations as part of an amplified ethical concern. Though there are already extant ethical values promulgated within the Standard Engineering enterprise, these must be interpreted and expanded to cover the global landscape. The Virtue Ethics values which are, for instance, the virtues of fairness, honesty, and care, and the Process Ethics values which are, for instance, environmental sustainability, social justice (which is becoming global justice), and health & safety must be brought into the assessment of the Global Engineering project.

4. A systems approach would be essential, especially a consideration of the world dynamics model of Jay Forester. The global system is fundamental, a constellation of Globalism and Globalization, a monumental mechanism spanning economic, cultural, and political systems. Woven throughout the program and woven into the global system would be ideas from courses in global economics, global philosophy, global politics, etc. A revival of some of the key ideas of the past ½ century of systems thinkers would admirably serve the Global Engineering program.

5. Last but not least, the Focal Engineering venture would provide the *ethos*, the characteristic spirit of the global engineering culture. Along with the Focal Engineering venture goes Material Ethics. It complements the above mentioned Virtue Ethics and Process Ethics, and it stresses the values of resonance, enlivenment, and engagement. In the global arena the focal product should ultimately be aimed at achieving the good by mitigating at least some problems of our global malaise, accomplishing at least some of the aims of critical theory, like advancing human emancipation from domination, exploitation, and injustice.

References

ABET. (2013). Criteria for Accrediting Engineering Programs. http://www.abet.org/DisplayTemplates/DocsHandbook.aspx?id=3149

Borgmann, A. (1993). *Crossing the postmodern divide*. Chicago: University of Chicago Press.

Brown, M. (2007). Globalism or globalization? *Modern Language Quarterly, 68*(2), 137–143.

El-Ojeili, C., & Hayden, P. (2006). *Critical theories of globalization*. Basingstoke/New York: Palgrave MacMillan.

Forester, J. (1971). *World dynamics*. Cambridge: Wright-Allen Press.

Held, D. (2004). Introduction to critical theory. In D. Rasmussen & J. Swindal (Eds.), *Critical theory: Volume I – historical perspectives*. London: Sage.

Imade, L. O. (2003). The two faces of globalization: Impoverishment or prosperity?. http://globalization.icaap.org/content/v3.1/01_imade.html

Jones, W. T. (1969). *Hobbes to Hume: A history of western philosophy* (2nd ed.). New York: Harcourt, Brace & World.

Kellner, D. (1996). Globalization and the postmodern turn. http://gseis.ucla.edu/faculty/kellner/essays/globalizationpostmodernturn.pdf

Khan, M. (2003). Teaching globalization, *The Globalist*. http://www.globalpolicy.org/globaliz/define/2003/0828teaching.htm

Kochler, H. (2000). Philosophical aspects of globalization. http://hanskoechler.com/rtg-hk.htm

Langhorne, R. (2001). *The coming of globalization*. Basingstoke/New York: Palgrave Macmillan.

Marcuse, H. (1973). *Studies in critical philosophy*. Boston: Beacon.

Moriarty, G. (2008). *The engineering project: Its nature, ethics, and promise*. University Park, PA: Penn State University Press.

Newberry, B. (2005). Engineering globalization: Oxymoron or opportunity? *IEEE Technology and Society Magazine, 24*(3), 8–15.

Nye, J. (2002). Globalism versus globalization. *The Globalist*. http://www.theglobalist.com/StoryId.aspx?StoryId=2392

Rifkin, J. (2009). *The empathic civilization: The race to global consciousness in a world in crisis*. Los Angeles: Tarcher.

Ritchie, M. (1996). Globalization vs. Globalism: Giving internationalism a bad name. http://www.itcilo.it/english/actrav/telearn/global/ilo/globe/kirsh.htm

Scholte, J. A. (2000). *Globalization: A critical introduction*. Basingstoke/New York: Palgrave Macmillan.

Index

© Springer International Publishing Switzerland 2015
C. Murphy et al. (eds.), *Engineering Ethics for a Globalized World*, Philosophy
of Engineering and Technology 22, DOI 10.1007/978-3-319-18260-5

CPSIA information can be obtained at www.ICGtesting.com
Printed in the USA
BVOW06*1308050715

407443BV00009B/103/P